ELECTRICIAN PRAYER MANUAL

SoulCraft Observer Effect

Chase DuQuesnay
EnQi ReaL

Amazon

Those who love Jesus Christ & Wusir...

CONTENTS

INTRODUCTION

The Perfect Black is the Flesh, Christ is the Light. You do not have to pretend that you hate Christians to study Ancient Egypt, you do not have to quote Racist, Inquisitionist Church Father form the Dark Ages regarding your potential Ancestors to be Christian.

Learning about Kemet, Angkor Wats, Sumer etc... does not stop you from studying the Bible. If they tell you Kemet, Angkor Wats or Sumer was brilliant and some or all of the Bible comes from those texts... The doesn't make the Bible smaller, it only makes the Bible bigger. There was a time when our Ancient civilizations were under attack for their knowledge, many scrolls, papyri and books were smuggled out of captivity in secret. What if the Bible is the Voltron of all those Ancient Sciences and Texts? I mean all the Bible Haters will tell you that right? Its Divination in there, it's Astrology in there, it's Gematria in there, it's

Masonry in there, its Gnostic texts in there, its just copies of Kemetic Science blah blah blah blah... The moment you realize all that stuff can arguably be found in one book, you gotta get into that book!

The moment you realize that NASA was started by a series of Satanic Rituals, you have to rethink science devoid of God. I have never been a fan of, nor have I ever taught science devoid of God. YOU JUST HAVE TO USE COMMON SENSE!!! Forget FlatEarth theories that is a plant.

You know that technology gets better with time, if we landed on the moon in 1969, how we aint been back? Did you even notice that the 50th anniversary was in 2019? Did Pauline Pierce really **know** Aleister Crowley? Does Crowley & Parsons have anything to do with Marconi Wireless Telegraph Company aka United Wireless Telegraph Company aka General Electric aka Federal Telegraph Company aka RCA? With his newly created Tesla coils, the inventor soon discovered that he could transmit and receive powerful radio signals when they were tuned to resonate at the same frequency. When a coil is tuned to a signal of a particular frequency, it literally magnifies the incoming electrical energy through resonant action. That company stole Tesla whole vibe and backrupted him, ran around teaching teaching and building of his I.P. Tesla went the legal route but they were all paid off, the U.S. Supreme Court upheld Tesla's radio patent number 645,576... after he died!!!

Your spine is a FM Radio antenna, I keep telling you this, prayer, radio, TV, Cell Phones and even space ships are using Radio Waves. If we landed on the moon, you may have missed the most important part of those moon landing stories. The most important science you overlooked was Radio! They supposedly communicated with the Asstronauts the whole time via Radio, the same radio waves we can't keep signals on 50+ years later.

Light and sound only exist in your mind. Space and Time only exist in your mind. We can experience them independently because all of our minds are linked to each other and to the WaveGuide. Why do you think the concept of Time and Space was so spooky to Einstein? Einstein was born to a Bible family and would have been classified as Autistic and began studying Masonry at **12**.

You dont see the trick? Every big 'Black' youtube platform sells Liquor! Hassan, Math, Tasha... We spoke about Spirits but do you know the origin of the word Liquor? If you want to hide something from 'Black People' where do you put it?

LQR - **means to read**, yes remember Ancient Languages didnt originally have vowels.

Do you understand a medium really is? Yes, a person that speaks to the living on behalf of the dead. NNNNNN!!! Wrong Answer! lmao... It's kinda that now, in the Bible though what is the medium for the spirit? The Flesh!

Welcome to Melanin vs Diabetes Ministry and Movement!

1 Corinthians 3:16-17

16 Know ye not that ye are the temple of God, and that the Spirit of God dwelleth in you?

17 If any man defile the temple of God, him shall God destroy; for the temple of God is holy, which temple ye are.

John 1:14

And the Word was made flesh, and dwelt among us, (and we beheld his glory, the glory as of the only begotten of the Father,) full of grace and truth.

Wait...

Let me ask something what if John 1 is true?

John 1: 1-8

1 In the beginning was the Word, and the Word was with God, and the Word was God.

2 The same was in the beginning with God.

3 All things were made by him; and without him was not any thing made that was made.

4 In him was life; and the life was the light of men.

5 And the light shineth in darkness; and the darkness comprehended it not.

6 There was a man sent from God, whose name was John.

7 The same came for a witness, to bear witness of the Light, that all men through him might believe.

8 He was not that Light, but was sent to bear witness of that Light.

What if space agencies are messing around with the Word, because the Word is God? That word being waves in radio wavelength pocket? If we look at a diagram of electromagnetic radiation, Radio waves would be the Alpha, just saying or maybe Omega depending on how we look at the diagram. No life can exist without Infrared and Infrared comes out of Radio Waves!

YOU HAVE TO HAVE ALL 3 OF THESE!!!
PRAYER BOOKS

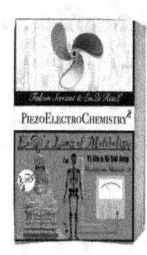

MYSTICAL VS SPIRITUAL

There is only 1 Most High, he is in heaven, I hope you read this book and ascribe all it's flaws to me and all the Light to him!

It has become super apparent that we, YOU, need this, based on the feed back to the prayer video. Lets walk through it… I am a Kemetic Student that identifies as a Messianic Jew, that identifies as a Nazarene (Christian).

What is that? Like what is a Christian? How do I become a Christian? Easy

Believe that you are loved and accepted by God.

Admit that you are a sinner.

Acknowledge your sins.

Commit your life to Christ.

Live a life of gratitude to God.

1

OK Doc so just say I believe in a man in the clouds? Bwahahahahaha That's so disrespectful, you think I believe in a man on top of the clouds? You probably need to get some of our other books first... Try the Speak it into Existence book maybe... Better yet get the PiezoElectroChemistry book! The bible does not say god is a man, at all. The bible says we are created in his image and likeness, it also says we are made from dirt with his breath. We have discussed that issue many times, your cells, your body is water and dirt. The human cell is the blueprint of life, it is in the image and likeness of earth, the sun, the stars etc... Selfishness starts right there in the mirror, thinking you _____. The does say that God is Light and if Im not mistaking his sun is too.

1 John 1:5 states that God is light - but Genesis 1:3 describes God creating light. Obviously God cannot create himself; nor does his nature change, right?

John 8:12

When Jesus spoke again to the people, he said, "I am the light of the world. Whoever follows me will never walk in darkness, but will have the light of life."

John 1

The Word Became Flesh

1 In the beginning was the Word, and the Word was with God, and the Word was God.
2 He was with God in the beginning.
3 Through him all things were made; without him nothing was made that has been made.
4 In him was life, and that life was the light of all mankind.

God is Music.

Ok if the Bible says God is Music, how do you proudly state God is Music? Light and Sound only exist in your mind! The Job of Neurons is to convert Light into Sound into arrangements of Adams (elements).

Arrangement - "act of arraigning, act of putting in proper order," 1740, from French arrangement (Old French arengement), from arranger "arrange" (see arrange). The meaning "that which is put in order, combination of parts or materials" is from 1800. The sense in music, "**adaptation of a composition to voices or instruments, or to a purpose, for which it has not been designed**," is by 1813. The meaning "final settlement, adjustment by agreement" is from 1855.

Arrange - late 14c., arengen, "draw up a line of battle," from Old French arengier "put in a row, put in battle order" (12c., Modern French arranger), from a- "to" (see ad-)

+ rangier "set in a row" (Modern French ranger), from rang "rank," from Frankish *hring or a similar Germanic source. , from Proto-Germanic *hringaz "**something curved, circle**," the source also of ring (n.1). It is reconstructed to be from a nasalized form of the PIE root *sker- (2) "to turn, bend."

It was a rare word until the meaning generalized to "to place things in order" c. 1780-1800. The sense of "come to an agreement or understanding" is by 1786. The musical sense of "adapt for other instruments or voices" is by 1808. Related: Arranged; arranging. Arranged marriage is attested by 1854.

Doesn't science tell you God is Music? Ok… Doesn't science tell you life arises from the Structured Water in your cells?

Structure - mid-15c.; the sense might be "**building materials**" or "**action or process of building or construction**," either way it is obsolete. From 1610s as "that which is built, an edifice," especially if large or imposing.

It is from Latin structura "a fitting together, adjustment; a building, mode of building;" figuratively, "**arrangement**, order," from structus, past participle of struere "to pile, place together, heap up; build, assemble, **arrange**, make by joining together," related to strues "heap" (from PIE *streu-, extended form of root *stere- "to spread").

It is attested from 1610s as "arrangement of parts," also "the frame or character of an organization." By late 17c. it was used in the broadest sense of "anything put together;" it is attested from 1746 in reference to literary works, by 1961 in linguistics.

13 - the number which brings the dead to life!!! Transition, movement, vibration etc... While the number 12 often symbolizes completeness (e.g., 12 tribes of Israel, 12 apostles), the addition of one more (13) may signify a movement towards chaos and transition away from that completeness. We will revisit this later...

Lets just recount how our society works.

We praise Killers aka Shooters on DemonTime.

We celebrate Homosexual Culture.

We promote hard Drug use to our Kids.

We frown on Church.

We totally violate Bible followers.

We complain that Catholics (Italians and Germans), Jewish people and Muslims own everything.

You don't see a problem with this because the religious people reaping God's blessings hire coons to mislead you. They put Oprah, Beyonce, Battlerappers, even Denzel up there to mislead you. Denzel done sold you Nagas that he performing gay rituals when he been married to one lady for 100 years smh... You watched Kevin Hart say he would not wear a dress, then he wore a dress, BOOOM Billionaire. Bwahahahaha... Nas said he would never sell liquor to his people... I mean they done put Conway & Pusha T in dresses, the 'toughest' nagas out, right?

We are in the age of the Fallen Stars, we now eat our kids! The movie, music and fashion industries are being shown to rape kids. You still avoiding the why? You think rich, sexy, handsome mean nd women need to drug and rape people for money? They got that already! Sex? They can get that anywhere. They are Satanist. Candace Owens said it to Nick Cannon and Nick lied his tail off... talking about he don't see it LMAO... He got them billions back, he is not allowed to speak the truth, he must support depopulation even after having 100 kids himself.

I been standing on this bible square holding my own for a loooong time...

You not religious but you spiritual? How?

You mean you worship the Devil right?

Bwahahahahaha

Mystical - late 15c., "enigmatic, obscure, symbolic," from **mystic** + **-al** (1). Meaning "having spiritual significance" is from 1520s. Related: Mystically.

late 14c., mistike, "**spiritually allegorical**, pertaining to mysteries of faith," from Old French mistique "mysterious, full of mystery" (14c.), or directly from Latin mysticus "mystical, mystic, **of secret rites**" (source also of Italian mistico, Spanish mistico), from Greek mystikos "secret, mystic, **connected with the mysteries**," from mystes "one who has been initiated" (see **mystery** (n.1)).
Meaning "pertaining to occult practices or ancient religions" is recorded by 1610s. That of "hidden from or **obscure to human knowledge** or comprehension" is by 1630s.

Mysticism - "**any mode of thought or life in which reliance is placed upon a spiritual illumination believed to transcend ordinary powers of understanding**," 1736, from **mystic** (adj.) + **-ism**. Often especially in a religious sense, and since the Enlightenment a term of reproach, implying self-delusion or dreamy confusion of thought.
Mysticism **and** rationalism **represent opposite poles of theology**, rationalism regarding the

reason as the highest faculty of man and the sole arbiter in all matters of religious doctrine; mysticism, on the other hand, **declaring that spiritual truth cannot be apprehended by the logical faculty**, nor adequately expressed in terms of the understanding. [Century Dictionary]

Spiritual - c. 1300, "of or concerning the spirit, **immaterial**" (especially in religious aspects), also "of or concerning the church," from Old French spirituel, esperituel (12c.) or directly from a Medieval Latin ecclesiastical use of Latin spiritualis "pertaining to spirit; of or pertaining to **breath, breathing**, wind, or air," from spiritus "of breathing; of the spirit" (see **spirit** (n.)).

The sense of "**originating with God**" is from late 14c. Related: Spiritually. An Old English word for "spiritual" was godcundlic. Spirital "pertaining to the spiritual realm" (from Latin spiritalis) also was in use from late 14c. to about 1700. Spirituose, a coinage of the 17c., was rare and now is obsolete.

In avibus intellige studia spiritualia, in animalibus exercitia corporalia [Richard of St. Victor (1110-1173): "**Watch birds to understand how spiritual things move**, animals to understand physical motion." - E.P.]

You not religious but you spiritual? How?

You mean you worship the Devil right? Maybe

you just mean you breathing? You mean you have a spirit huh? You can't be spiritual if you don't believe in God!

God is Music.

Nature is his Producer.

Pigments are his Beat Machine.

Mystic study is Abstract study, your brain is hardwired for the study of God. Spiritual is immaterial, the opposite of and precursor to material. You were informed to watch birds to see how spiritual things moved... Birds demonstrate the nature of particles and waves, close up it's separate individual birds/particles. When you see birds from a distance, the flock moves like water or a wave. Birds show you how drops of water make a ocean.

Sidebar: The amygdala is a part of a network that modulates respiration, and it is the network response that dictates whether respiration is increased or decreased in response to amygdala activation or engagement.

Forebrain control of breathing: Anatomy and potential functions

Karl M. Schottelkotte[1]   Steven A. Crone[2,3,4*]

The forebrain plays important roles in many

critical functions, including the control of breathing. We propose that the forebrain is important for ensuring that breathing matches current and anticipated behavioral, emotional, and physiological needs. This review will summarize anatomical and functional evidence implicating forebrain regions in the control of breathing. These regions include the cerebral cortex, extended amygdala, hippocampus, hypothalamus, and thalamus. We will also point out areas where additional research is needed to better understand the specific roles of forebrain regions in the control of breathing.

Introduction

Breathing is an essential function for humans during every waking and sleeping moment to provide movement of air in and out of the lungs (1–3). Our oxygen demands are dynamic and constantly changing depending on our activity level, emotional state, health status, and current behaviors. Because of this, the brain must constantly ensure that our breathing is appropriately matched to our physiological state and behavior. For example, our breathing rate and/or volume changes in anticipation of altered needs for gas exchange and tissue oxygenation during exercise or other physical activities (4). This feed-forward control is critical for maintaining homeostasis because there is no known mechanism for sensing gas exchange in the muscle or lungs. Breathing changes with emotional states as well, as the feelings of stress and fear can cause hyperventilation (5–7). Although the basic pattern of respiration (inspiration, post-inspiration, and expiration phases) is generated by neurons in the brainstem and transmitted to respiratory muscles via spinal circuits (1, 2, 8), these neurons are influenced by other brain regions in order to ensure that breathing is appropriate for the current situation. Breathing, unlike other autonomic processes such as heart rate and blood pressure, can be modulated voluntarily in addition to autonomically (9–11). For example, singers and musicians that play wind instruments need precise control over their breathing to produce the correct notes and tones. Mindfulness exercises such as meditation and yoga utilize deliberate and precise breathing methods to elicit calming responses from the body, including lowering blood pressure and heart rate. Competitive weightlifters are among a variety of athletes that use methodic breathing techniques such as hyperventilating before their lift to provide sympathetic activation to maximize strength during their lift. Swimmers pace their breathing to ensure that they do not inadvertently inhale water. Thus, intentionally pacing respiration or modulating breathing volume is a tool that humans and animals use to control their own physiology.

The forebrain is important for the planning

and execution of movements, sensory processing, regulating sleep wake states and behavioral responses to emotions such as stress and fear (5, 7). Each of these functions can have an impact on breathing. For example, fear is linked with a variety of respiratory changes- you may gasp if you are startled, you might find yourself holding your breath when scared, or even hyperventilating as your body prepares its fight or flight response. Sleep/wake states strongly influence the control of breathing, with important consequences if this relationship is dysfunctional, such as sleep apnea, congenital central hypoventilation syndrome, sudden infant death syndrome, or sudden unexpected death in epilepsy. Thus, we propose that the forebrain may be important for ensuring that breathing matches current and anticipated emotional, behavioral, and physiological needs. However, the circuits and mechanisms by which the forebrain exerts control over breathing are only partly understood. This review will summarize anatomical and functional evidence implicating forebrain regions in the control of breathing. These regions include the cerebral cortex, extended amygdala, hippocampus, hypothalamus, and thalamus (Figure 1). We will also point out areas where additional research is needed to better understand the specific roles of forebrain regions in the control of breathing.

You not religious but you spiritual? How?

Do you see why I have to 'come out the closet'? My God, thee God, has been blessing since I was a boy. I have always been shielded and protected, then I have healed thousands, those who I have taught have done the same. God has literally healed millions around the world through me, many women couldn't have children and we changed that! There are humans walking the planet that would not be here if not for me, my relationship with him, the support from you! I am your profit, not prophet but profit. If you have spent time, paid attention, balanced your interests accordingly, you should've had a great ROI. My spirit leads me, I read:

Lamentations 2:11

My eyes are filled with tears. My stomach is in torment. My Liver is poured out on the ground over the destruction of the daughter of my people — as young children and infants languish in the city squares.

After the last book can you see how this would peek my curiosity?

I immediately go hmm let me see what the older papers say, sure enough I notice that my name sake Kheper may very well be the source for the liver. Keep in mind always, Greece was educated by

Kemet.

Liver - **secreting organ of the body**, Old English **lifer**, from Proto-Germanic *librn (source also of Old Norse <u>lifr</u>, Old Frisian livere, Middle Dutch levere, Dutch lever, Old High German lebara, German Leber "**liver**"), perhaps from PIE root <u>*leip-</u> "to stick, adhere," also used to form words for "<u>**fat**</u>."
Formerly believed to be the body's blood-producing organ; <u>**in medieval times it rivaled the heart**</u> as the supposed seat of love and passion. Hence lily-livered, a white (that is, bloodless) liver being supposed a sign of cowardice, Shakespeare's pigeon-livered, etc. Liver-spots, once thought to be caused by a dysfunction of the organ, is attested from 1730. Greek **hépat**-, s. of **hêpar** liver + New Latin –cyta, from Greek kutos, hollow vessel.

The trilateral root for Kheper is Hpr, the only organ the regenerates just so happens to be named after the God of Regeneration. I guess that is a co-inky dink? The Beetle rolls its eggs in poop while the human semen is controlled by the Amygdala & Liver! I learn this from the Bible which sends me further back to Kemet, I then go search the journals to see if I can substantiate such a wild claim.

Amygdala-liver signaling orchestrates rapid glycemic responses to stress and drives

stress-induced metabolic dysfunction

Sarah Stanley 1, Kavya Devarakonda 1, Richard O'Connor 1, Maria Jimenez-Gonzalez 1, Alexandra Alvarsson 1, Rollie Hampton, Diego Espinoza 1, Rosemary Li 1, Abigail Shtekler 1, Kaetlyn Conner 1, Mitchell Bayne 1, Darline Garibay 2, Jessie Martin 1, Vanessa Lehmann 1, Liheng Wang 1, Paul Kenny 1

Abstract

Behavioral adaptations to environmental threats are crucial for survival and necessitate rapid deployment of energy reserves. The amygdala coordinates behavioral adaptations to threats, but little is known about its involvement in underpinning metabolic adaptations. Here, we show that acute stress activates medial amygdala (MeA) neurons that innervate the ventromedial hypothalamus (MeAVMH neurons), which precipitates hyperglycemia and hypophagia. The glycemic actions of MeAVMH neurons occur independent of adrenal or pancreatic glucoregulatory hormones. Instead, using whole-body virus tracing, we identify a polysynaptic connection from MeA to the liver, which promotes the rapid synthesis of glucose by hepatic gluconeogenesis. Repeated stress exposure disrupts MeA control of blood glucose and appetite, resulting in diabetes-like dysregulation of glucose homeostasis and weight gain. Our findings reveal a novel amygdala-liver axis that regulates rapid glycemic adaptations to stress and links recurrent stress to metabolic dysfunction.

Nonalcoholic fatty liver disease and alteration in semen quality and reproductive hormones

Yan Li 1, Lei Liu, Bin Wang, Dongfeng Chen, Jun Wang

Abstract

Objectives: Nonalcoholic fatty liver disease (NAFLD) is the most common cause of chronic liver disease in the world. Some reports have shown that NAFLD may cause multisystem damage, but its influence on male reproductive function has rarely been studied.

Aim: To evaluate the influence of NAFLD on sperm quality and reproductive hormones in Chinese men.

Materials and methods: A total of 102 NAFLD men and 94 healthy men without fatty liver (control) were enrolled in this study. All participants underwent a physical examination, and were subjected to lifestyle questionnaires and abdominal ultrasound examination. The semen quality (volume, concentration, motility, and morphology) and serum hormonal levels (testosterone, estradiol, follicle-stimulating hormone, inhibin B, sex hormone-binding globulin, and luteinizing hormone) were examined and compared between the two groups.

Results: The levels of serum testosterone and sex hormone-binding globulin were significantly lower in the NAFLD patients compared with the control group. Sperm concentration (P=0.04), sperm count (P=0.01), and total motility (P=0.03) in the NAFLD patients were significantly decreased compared with the control group. However, no significant differences were observed in semen volume and morphology. Multivariate analysis showed that sperm concentration, sperm count, and motility were significantly associated with NAFLD and abstinence (P<0.05 or P<0.001).

Conclusion: These results suggest that NAFLD could significantly affect sperm quality and reproductive hormones.

Lamentations 2:11

My eyes are filled with tears. My stomach is in torment. My Liver is poured out on the ground over the destruction of the daughter of my people — as young children and infants languish in the city squares.

My Liver is poured out on the ground over the destruction of the daughter of my people... that part! That's a thing that will get my spidey senses tingling and … that curiosity doesn't always pan out lol… Most times though we find gold!

YOU HAVE TO HAVE ALL 3 OF THESE!!!
PRAYER BOOKS

THE SKYFI

What is the SkyFi, the WaveGuide? Your Lord and Shepard.

I get it, this may be a lot all at once but you follow Rhesus or Jesus, your choice? There is a reason they are making a full court press pushing the homosexuality and low grade beastiality. Watch for the signs, pay attention to the numbers around you. Keep in mind number not only describes music, it is a word that means a song.

Number - c. 1300, "**sum**, aggregate of a collection," from Anglo-French noumbre, Old French nombre and directly from Latin numerus "a number, quantity," from PIE root *nem- "**assign**, allot; take."

The meaning "written symbol or **figure of arithmetic value**" is from late 14c. The meaning "single (numbered) issue of a magazine" is from 1795. The colloquial sense of "**a person or thing**" is by 1894. The meaning "dialing combination to reach a particular telephone receiver" is from

1879; hence wrong number (1886).

The sense of "**musical selection**" (1885) is from popular theater programs, where acts were marked by a number. Earlier numbers meant "**metrical sound or utterance, measured or harmonic expression**" (late 15c.) and, from 1580s, "**poetical measure, poetry, verse.**"

Number one "oneself" is from 1704 (mock-Italian form numero uno attested from 1973); the biblical Book of Numbers (c. 1400, Latin Numeri, Greek Arithmoi) is so called because it begins with a census of the Israelites. Childish slang number one and number two for "urination" and "defecation" attested from 1902. Number cruncher is 1966, of machines; 1971 of persons.

To get or have (someone's) number "have someone figured out" is attested from 1853; to say one's number is up (1806) meaning "one's time has come" is a reference to the numbers on a lottery, draft, etc. The numbers "illegal lottery" is from 1897, American English. Do a number on is by 1969, exact meaning unclear; by the early 1970s it can mean "emotionally manipulate" (1970), "damage or injure" (1975), or "assassinate, kill" (1971). The 1972 book of gay slang The Queen's Vernacular says it is synonymous with game, as well as with trick in the prostitution and magical senses, and defines it as "one's skit, act, schtick; contrived actions used to gain attention." The image may be of a routine song-and-dance performance, which if so makes it from the "**musical selection**" sense.

c. 1300, "to count," from Old French nombrer "to count, reckon," from nombre (n.) "number" (see number (n.)). Meaning "to assign a distinctive number to" is late 14c.; that of "**to ascertain the number of**" is from early 15c. Related: Numbered; numbering.

Quantum - 1610s, "**sum, amount**," from Latin quantum (plural quanta) "as much as, so much as; **how much**? how far? how great an extent?" neuter singular of correlative pronominal adjective quantus "as much" (see quantity).
The word was introduced in physics directly from Latin by Max Planck, 1900, on the notion of "**minimum amount of a quantity** which can exist;" reinforced by Einstein, 1905. Quantum theory is from 1912; quantum mechanics, 1922. The term quantum jump "abrupt transition from one stationary state to another" is recorded by 1954; quantum leap "sudden large advance" (1963), is often figurative.

Probability - mid-15c., probabilite, "**likelihood of being realized**, appearance of truth, quality of being probable," from Old French probabilite (14c.) and directly from

Latin probabilitatem (nominative probabilitas) "**credibility**, probability," from probabilis (see probable).

Meaning "something likely to be true" is from 1570s; mathematical sense is from 1718, "**frequency** with which **a proposition** similar to the one in question is **found true** in the course of experience."

In weather forecasting, probabilities was used in U.S. from 1869 and adopted in the official weather forecasts of the United States Signal Service; hence Old Probabilities, a humorous name for the chief signal officer of the Signal Service Bureau (by 1875).

Quantum = Math

God is Music

Ahmes says "Accurate reckoning. **The entrance into the knowledge of all existing things and all obscure secrets**. This book was copied in the year 33, in the fourth month of the inundation season, under the majesty of the king of Upper and Lower Egypt, 'A-user-Re', endowed with life, in likeness to writings of old made in the time of the king of Upper and Lower Egypt, Ne-ma'et-Re'. It is the scribe Ahmes who copies this writing".

Do you really understand the Rhind Mathematical Papyrus? Do you get the Eye of Heru now?

Observance through measurement…?

Observance is participation. Think about all the people that film police brutality, or watch and share it endlessly, but don't participate in local politics. I will leave that alone, let's deal with the science at hand (pun intended).

Faith is Observance.

I know you may be a little lost, we are discussing the cover for Spirituality here. The Double Slit experiment and the Observer Effect.

Faith with Acts is Dead.

Observing is Participating.

Observing/Participating creates a probability amplitude at the quantum level. The extent of your observance increase this amplitude of Wave Height. This is how you create Order from Chaos, chaos is simply pure energy going **to and fro**. The magick is in here, observe is not only participating it is a measurement. Measuring what? Probability (Fervency & Zeal). Poof! The observer is now entangled in that which he/she is observing, participating in the outcome. This is pure a aspect of the soul, divine light, so matter doesn't matter.

Without the constraints of matter, time and distance are not relevant here. Welcome to spooky action at a distance, wait let me add a caveat.

Power and Voltage do matter (pun intended lol). This is why quantum science is taught in colleges around the world using Chlorophyll and Melanin, pigment is the only natural biological material that facilitates Quantum action, Spooky!

Wait... You know Spooky mean Black? Bwahahahahaha...

Point is this is Soul Power, so Quantum Entanglement is a nonlocal phenomena. Objects become link and direct influence each other instantly, you and are I doing it right now! On a very spooky level your DNA is a wave probability magic wand. The environment and your voltage, activate or deactivate... We have a entire book on transcription and translation, please read &/or reread our Declaration of Independence book.

Before we get to the Prophet Ezekiel's description of the WaveGuide and all it's splendor, aka the SkyFi. I want to just define Wifi...

Wifi - Wi-Fi is a wireless networking technology that uses radio waves to provide wireless high-speed Internet access.

Internet - a vast network that connects computers all over the world. Through the Internet, people can share information and communicate from anywhere with an Internet connection.

Computer - 1640s, "**one who calculates**,

a reckoner, **one whose occupation is to make arithmetical calculations**," agent noun from compute (v.).

Meaning "**calculating machine**" (of any type) is from **1897**; in modern use, "programmable digital electronic device for performing mathematical or logical operations," 1945 under this name (the thing itself was described by 1937 in a theoretical sense as Turing machine). ENIAC (1946) usually is considered the first.

Computer literacy is recorded from 1970; an attempt to establish computerate (adjective, on model of literate) in this sense in the early 1980s didn't catch on. Computerese "the jargon of programmers" is from 1960, as are computerize and computerization.

WASHINGTON (AP) — A New York Congressman says the use of computers to record personal data on individuals, such as their credit background, "is just frightening to me." [news article, March 17, 1968]

Earlier words for "one who calculates" include computator (c. 1600), from Latin computator; computist (late 14c.) "one skilled in calendrical or chronological reckoning."

1897 - the year that a computer became known as a machine instead of human, The Scientific-Humanitarian Committee (Wissenschaftlichhumanitäres Komitee, WhK) is founded in Berlin as an LGBT campaigning organization, the first such in history, Bram Stoker's contemporary Gothic horror novel Dracula is first published (in London), J. J. Thomson first describes his discovery of the electron, Guglielmo Marconi sends the first ever wireless communication over open sea when the message "Are you ready" is transmitted across the Bristol Channel from Lavernock Point in South Wales to Flat Holm Island, a distance of 6 kilometres (3.7 mi), Thomas Edison is granted a patent for the Kinetoscope, a precursor of the movie projector... That's a Hell of year (pun intended)! Before you say anything, I am going to sure, all these events are co-incidentally just happening at the same time.

Just for sh!ts and giggles I am going to add some other little tidbits here...

Helena Petrovna Blavatsky[a] (née Hahn von Rottenstern; 12 August [O.S. 31 July] 1831 – 8 May 1891), often known as Madame Blavatsky, was a Russian and American mystic and author who co-founded the Theosophical Society in 1875. She gained an international following as the primary founder of Theosophy as a belief system. This is the system based on Hermeticism, which has given rise to the **Hermaphrodite God**. Aleister Crowley (/ˈælɪstər ˈkroʊli/ AL-ist-ər KROH-lee; born Edward

Alexander Crowley; 12 October 1875 – 1 December 1947) was an English occultist, ceremonial magician, poet, philosopher, political theorist, novelist, mountaineer, and painter. He founded the religion of Thelema, identifying himself as the prophet entrusted with guiding humanity into the Æon of Horus in the early 20th century. A prolific writer, he published widely over the course of his life.

Born to a wealthy family in Royal Leamington Spa, Warwickshire, Crowley rejected his parents' fundamentalist Christian Plymouth Brethren faith to pursue an interest in Western esotericism. He was educated at Trinity College at the University of Cambridge, where he focused his attention upon mountaineering and poetry, resulting in several publications. Some biographers allege that here he was recruited into a British intelligence agency, further suggesting that he remained a spy throughout his life. In 1898, he joined the esoteric Hermetic Order of the Golden Dawn, where he was trained in ceremonial magic by Samuel Liddell MacGregor Mathers and Allan Bennett. Just know that there is codewords associated with these people. The Emerald Tablets are their replacement for Kemetic-Biblical thought. Th Emerald Tablets have nothing to do with Djehuti of Egyptian origin, nor are they even related to Thoth as they claim! They are a modernization of Greek Man-Boy sex cults codified with some stolen knowledge, reverse engineered for Satanism (ill explain that very simply later). There is nothing godly in these groups and **THEY ARE NOT MASONS**!!! The Democrats are into this, the Abortion Clinics are chopping up the Babies and selling the body parts, Satanist just buy dead babies now like a super market!

Crowley even wrote 'the Book of the Law' to completely replace all Holy Scriptures. The original manuscript was sent on Crowley's death to Karl Germer, the executor of his will and head of_Ordo Templi Orientis (O.T.O.). This is what is going on when the "lights go out" and your unconscious with these folks.. You forgot that girl Polight is in jail for said he gave her a bunch of drugs and she passed out... Yall better wake up! It's a war going on outside no man is safe from, you can run but you can't hide forever, from these streets that we done took, you walking with ya head down scared to look (observe), you shook, cause aint no such thing as halfway crooks, they never around when the beef cooks... - Albert Johnson aka Prodigy Rest in Greatness! Back to live action.

Wifi - Wi-Fi is a wireless networking technology that uses radio waves to provide wireless high-speed Internet access.

Internet - a vast network that connects computers

all over the world. Through the Internet, people can share information and communicate from anywhere with an Internet connection.

We now know the original computer was a person, right? The WaveGuide and your central nervous system is a much much more powerful internet, but everyone didn't have access and amongst those that do, their is a hierarchy. Please read &/or reread the Speak it into Existence book!

Computer Network - A computer network is a system that connects two or more computing devices for transmitting and sharing information. Computing devices include everything from a person, mobile phone to a **server**. These devices are connected using physical wires such as fiber optics, but they can also be **wireless**. These interconnections are made up of telecommunication network technologies based on physically wired, optical, and wireless **radio frequencies**.

Network Server - a computer system that provides services, data, applications, or network management to other computers, known as **clients**, over a network. The primary role of a network server is to manage and facilitate communication and resource sharing amongst networked computers.

Client - late 14c., "**one who lives**

under the patronage of another," from Anglo-French clyent (c. 1300), from Latin clientem (nominative cliens) "**follower**, retainer" (related to clinare "to incline, bend"), from PIE *klient-, a suffixed (active participle) form of root *klei- "to lean." The notion apparently is "one who leans on another for protection." In ancient Rome, a plebeian under the guardianship and protection of a patrician (who was called patronus in this relationship; see **patron**).

The meaning "a lawyer's customer" is attested from c. 1400, and by c. 1600 the word was extended to any customer who puts a particular interest in the care and management of another. Related: Cliency.

The relation of client and patron between a plebeian and a patrician, although at first strictly voluntary, was hereditary, the former bearing the family name of the latter, and performing various services for him and his family both in peace and war, in return for advice and support in respect to private rights and interests. Foreigners in Rome, and even allied or subject states and cities, were often clients of Roman patricians selected by them as patrons. The number of a patrician's clients, as of a baron's vassals in the middle ages, was a gage his greatness. [Century Dictionary]

Observer = Influencer

Client = Followers/Subs

This is basically a teacher student relationship, the start of all massive world changing movements. To stop such a movement from developing though, they created a synthetic medium where all future relationships can be analyzed, enhanced or destabilized. It is very important that in this new world we transmit information offline! Wakey Wakey!!!

Now let check the Big Homey Ezekiel description of the ya Digg..

Ezekiel

1 Now it came to pass in the thirtieth year, in the fourth month, in the fifth day of the month, as I was among the captives by the river of Chebar, that the heavens were opened, and I saw visions of God.

2 In the fifth day of the month, which was the fifth year of king Jehoiachin's captivity,

3 The word of the Lord came expressly unto Ezekiel the priest, the son of Buzi, in the land of the Chaldeans by the river Chebar; and the hand of the Lord was there upon him.

4 And I looked, and, behold, a whirlwind came out of the north, a great cloud, and a fire infolding itself, and a brightness was about it, and out of the midst thereof as the colour of amber, out of the midst of the fire.

5 Also out of the midst thereof came the likeness of four living creatures. And this was their appearance; they had the likeness of a man.

6 And every one had four faces, and every one had four wings.

7 And their feet were straight feet; and the sole of their feet was like the sole of a calf's foot: and they sparkled like the colour of burnished brass.

8 And they had the hands of a man under their wings on their four sides; and they four had their faces and their wings.

9 Their wings were joined one to another; they turned not when they went; they went every one straight forward.

10 As for the likeness of their faces, they four had the face of a man, and the face of a lion, on the right side: and they four had the face of an ox on the left side; they four also had the face of an eagle.

11 Thus were their faces: and their wings were stretched upward; two wings of every one were joined one to another, and two covered their bodies.

12 And they went every one straight forward: whither the spirit was to go, they went; and they turned not when they went.

13 As for the likeness of the living creatures, their appearance was like burning coals of fire, and like the appearance of lamps: it went up and down among the living creatures; and the fire was bright, and out of the fire went forth lightning.

14 And the living creatures ran and returned as the appearance of a flash of lightning.

15 Now as I beheld the living creatures, behold one wheel upon the earth by the living creatures, with his four faces.

16 The appearance of the wheels and their work was like unto the colour of a beryl: and they four had one likeness: and their appearance and their work was as it were a wheel in the middle of a wheel.

17 When they went, they went upon their four sides: and they turned not when they went.

18 As for their rings, they were so high that they were dreadful; and their rings were full of eyes round about them four.

19 And when the living creatures went, the wheels went by them: and when the living creatures were lifted up from the earth, the wheels were lifted up.

20 Whithersoever the spirit was to go, they went,

thither was their spirit to go; and the wheels were lifted up over against them: for the spirit of the living creature was in the wheels.

21 When those went, these went; and when those stood, these stood; and when those were lifted up from the earth, the wheels were lifted up over against them: for the spirit of the living creature was in the wheels.

22 And the likeness of the firmament upon the heads of the living creature was as the colour of the terrible crystal, stretched forth over their heads above.

23 And under the firmament were their wings straight, the one toward the other: every one had two, which covered on this side, and every one had two, which covered on that side, their bodies.

24 And when they went, I heard the noise of their wings, like the noise of great waters, as the voice of the Almighty, the voice of speech, as the noise of an host: when they stood, they let down their wings.

25 And there was a voice from the firmament that was over their heads, when they stood, and had let down their wings.

26 And above the firmament that was over their heads was the likeness of a throne, as the appearance of a sapphire stone: and upon the likeness of the throne was the likeness as the appearance of a man above upon it.

27 And I saw as the colour of amber, as the appearance of fire round about within it, from the appearance of his loins even upward, and from the appearance of his loins even downward, I saw as it were the appearance of fire, and it had brightness round about.

28 As the appearance of the bow that is in the cloud in the day of rain, so was the appearance of the brightness round about. This was the appearance of the likeness of the glory of the Lord. And when I saw it, I fell upon my face, and I heard a voice of one that spake.

Dyson Spheres - A Dyson sphere is a hypothetical megastructure that encompasses a star and captures a large percentage of its power output. The concept is a thought experiment that attempts to imagine how a spacefaring civilization would meet its energy requirements once those requirements exceed what can be generated from the home planet's resources alone. Because only a tiny fraction of a star's energy emissions reaches the surface of any orbiting planet, building structures encircling a star would enable a civilization to harvest far more energy.

The umbilical chord of the Sun to the Earth, create the magnetosphere. The magnetosphere has its own movements, it surrounds everything. Within that sphere is the waveguide. Spheres within spheres...

Egypt described it so Nasa just decoding and remixing... Let see what the Oldheads in Egypt was   on....

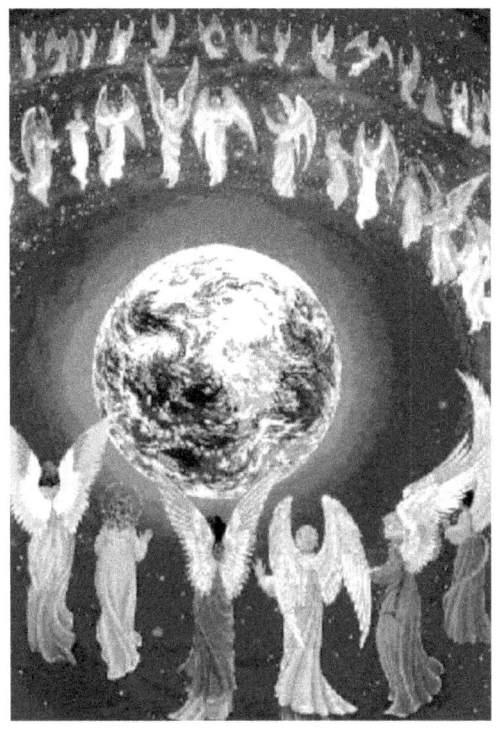

Anyone in 'the Lodge' should recognize these things quickly...

Coffin Text 76 "Chambers of the Sky"
Below is Aat VIII from the Papyrus of Nu

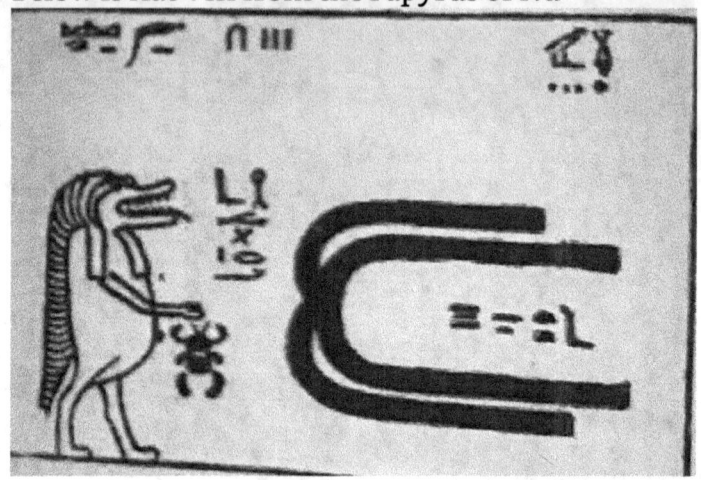

Sidebar: This is why I had to bring the waves back... Nas selling liquor now, he can't rep the crown... We have to take this uptown, back over the WhiteStone!Jesus is spelled with a J but pronounced with a G, a circled 7. You know what my bad this is extra...

Focus!

Look at the similarity between what the Kemetic Scribes drew and our scientific idea, of the 'Duat'.

Remarkable!

I mean, to create that blind, we are assuming they had no telescopes etc... This wouldn't be the first time though, the Dogon were describing so serious cosmic phenomena without any tools too, right?

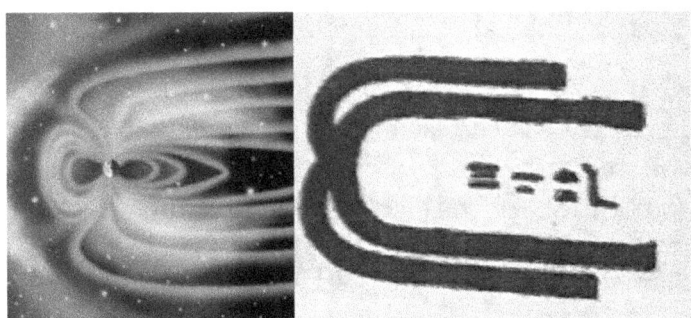

Coffin Text 474, 343, 344 & Pyramid Texts 214 'Fish Traps of Re' - Iron floats in the sky and iron weights on earth!

Coffin Text 62 District of the Waters with Ropes of Iron

The Pyramid Texts 509 get into the WaveGuide/ Shu, connecting them referring to the magnetosphere as "Ramparts of Shu"...

Rampart - "earthen elevation around a place for fortification," capable of resisting cannon shot and sometimes also including parapets, 1580s, from French rempart, rampart, from remparer "to fortify," from re- "again" (see re-) + emparer "fortify, take possession of," from Old Provençal amparer, from Vulgar Latin *anteparare "prepare," properly "to make preparations beforehand," from Latin ante "before" (from PIE root *ant- "front, forehead," with derivatives meaning "in front of, before") + parare "to get, prepare" (from PIE root *pere- (1) "to produce, procure"). With unetymological -t in French, perhaps by influence of boulevart (see boulevard).

Let's see if the scientific journals agree that the magnetosphere or 'fish trap of Re' is the foundation for the electricity in the air (SHU).

Solar Influence on Lightning Rates

The relationship between the magnetosphere and lightning involves a complex interplay of solar activity and atmospheric phenomena. Research has shown that the Sun's magnetic field has a significant impact on the occurrence of lightning on Earth. Since the 1930s, scientists have proposed that solar activity influences the frequency of lightning strikes. Recent studies have demonstrated that variations in the Sun's magnetic field can affect lightning rates not only over long timeframes but also on a daily basis. For example, during periods when the Sun's magnetic field points towards Earth, the UK experienced 50% more thunderstorms and subsequent lightning strikes compared to when it pointed away.

Mechanisms of Interaction

Several mechanisms have been identified that explain how the magnetosphere influences lightning. One significant finding is related to the heliospheric current sheet (HCS), a structure formed by the Sun's magnetic field. As the HCS crosses Earth, changes in lightning rates are observed within a short timeframe. Specifically, after certain switches in the magnetic polarity of the Sun's field, thunderstorms tend to peak, leading to increased lightning strikes shortly thereafter.

Another mechanism involves the interaction of cosmic rays with thunderclouds, which is influenced by shifts in the magnetosphere. When the Earth's magnetic field is altered, it allows cosmic particles to penetrate the atmosphere, potentially providing the energy needed to ignite lightning in pre-charged clouds.

Summary of Effects

The effects of the magnetosphere on lightning can be summarized as follows:

Magnetic Field Changes: Shifts in the Sun's magnetic field can lead to increased lightning rates by altering atmospheric electric circuits.

Cosmic Particles: Variability in the magnetosphere can enable cosmic rays to reach the atmosphere, affecting the likelihood of lightning in charged clouds.

Predictive Potential: Understanding these interactions could enhance forecasting models for lightning, integrating terrestrial weather patterns with solar activity predictions.

These findings indicate a fascinating connection between celestial phenomena and terrestrial weather, highlighting the Sun's influential role in shaping our atmosphere and **electrifying thunderclouds**.

Ezekiel was/is all wise, right and exact. The ancient spooks from egypt were also all wise, right and exact. Now this NASA Dyson stuff is fraud, they love reading our ancient papers and then saying they made up this or invented that.... Anyway God is the Original Server, he is at your service, ask and you shall receive.

Matthew 7:7: "Ask, and it will be given to you; seek, and you will find; knock, and it will be opened to you".

Mark 11:24: "Therefore I tell you, whatever you ask in prayer, believe that you have received it, and it will be yours".

John 14:13-14: "Whatever you ask in my name, this I will do, that the Father may be glorified in the Son. If you ask me anything in my name, I will do it".

James 4:3: "You ask and do not receive, because you ask with wrong motives, so that you may spend it on your pleasures".

YOU HAVE TO HAVE ALL 3 OF THESE!!! PRAYER BOOKS

CIRCLE OF ENERGY

Now that we are getting use to the idea that every thought, step, bite, word, every everything we say or do is living in the environment, circling us at light speed... We have discuss the speed of light as traveling around earth 7 times a second (circle 7/ G). The magnificent 7th Letter G in all it's splendor

 below:

The Akkadian word for house is beyth or betu, which is the same as the Arabic beyt or the Hebrew beyt, he explains. Dictionary of Akkadian Language Links Modern Civilization with Ancient

Origins 2011. We flipped the house of God around left to right and added the 7.

Remember in the Speak it into Existence book...

What is the co-inky dink Doc? The Rosary has around it's main part or body 54 (5+4=9) beads. It also has 5 sections divided by 4 beads.... I'm going to just leave you some abstract study, you can't really complete in a few clicks..

Go ahead and google 'secret of rosary' or "54 beads'.... You will find nothing connected to what I am talking about, you will learn about a rosary and grown some Brain/Heart cells though.

Here's a hint Jesus **rose** in how many hours? 72 right... Hapi fishing! Becareful going to prayer in a bad emotional state, your emotional state can betray you, your mouth may ask for one thing but you heart full of anger or jealousy may ask for something else.

The number of beads varies by religion or use. Islamic prayer beads, called Misbaha or Tasbih, usually have 100 beads (99 +1 = 100 beads in total or 33 beads read thrice and +1). Buddhists and Hindus use the Japa Mala, which usually has 108 beads, or 27 which are counted four times. Bahá'í prayer beads consist of either 95 beads or 19 beads, which are strung with the addition of five beads below. The Sikh Mala also has 108 beads. With the exception of 19 which we discussed in the Horus and Set TransAction book... All of these prayer beads are multiples of the number 9.

The I Ching, or "Yijing" (易經), often translated as the "Book of Changes" or "Classic of Changes," is an ancient Chinese divination text that is among the oldest of the Chinese classics. Its rich history, philosophical significance, and usage in divination practices make it an important cultural and intellectual artifact.

Historical Background

The I Ching has been traditionally dated to the **late 9th century BC.** The EyeKing is dated to the 9th century Bwahahahahaha...

At its core, the I Ching consists of **64** hexagrams, which are figures composed of six stacked horizontal lines (爻, yáo). Each line can be broken (yin) or solid (yang), determining the hexagram's meaning. The text features both hexagram statements and line statements that provide interpretations. The hexagrams are traditionally arranged according to the King Wen sequence, which pairs hexagrams with their upside-down equivalents, revealing deeper insights when interpreted as dynamic representations of change. This is system of binary code group by the numbers 3 and 6. The numbers 3 and 6 create

Vortex Math.

1 + 2 = 3

2 + 4 = 6

4 + 8 = 12 = 3

8 + 7 = 15 = 6

7 + 5 = 12 = 3

5 + 1 = 6

1 + 2 = 3

1
2
4
8
7
5
R
E
P
E
A
T

The pattern of movement is 1 - 2 - 4 - 8 - 7 - 5

The master 6, the number of man

6 coordinates or "angles"

3 + 6 = 9

Vortex Math is characterized by a system of numbers that form specific patterns, particularly focusing on the significance of the number 9. This number is described as central to the structure of the math, serving as an axis around which other numbers rotate. The approach typically involves reducing numbers to their digital roots, allowing complex operations to be simplified into single-digit representations.

One of the key sequences in Vortex Math is the pattern derived from doubling numbers and reducing them to single digits. The repeating cycle identified is 1, 2, 4, 8, 7, and 5, creating a visual representation often likened to a vortex or infinity symbol. This numerical arrangement suggests an inherent harmony and balance, which proponents claim reflects energy flow in the universe. It could just be the energy flow here, 7 times a second!

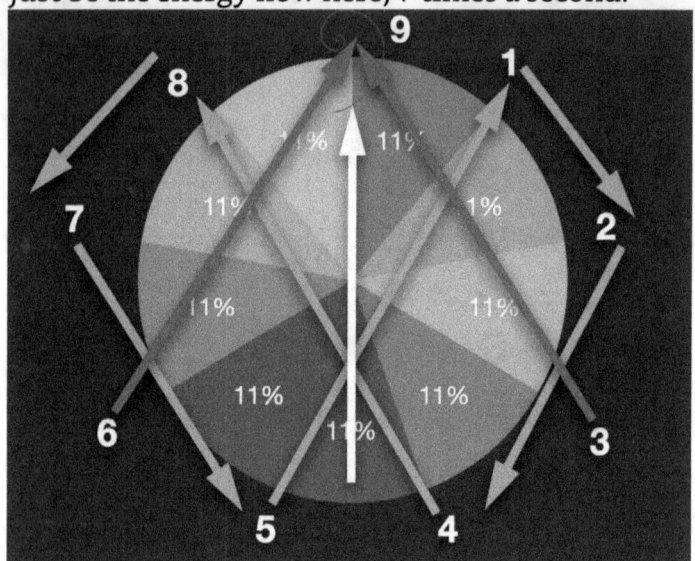

If you notice there is a 'Pyramid' hidden in the center of the energy flowing. There is also the fact that when you overlay this on a circle, that 9/11 pop out! There is also two big rectangles or oblong squares there... I leave that there...

Mind you Jesus rose after 72 hours right?

Wait check this, Im going to give you two versions of the same scripture and then move on...

Job 40:21-22

"Under the lotus plants he lies, in the shelter of the reeds and in the marsh. For his shade the lotus trees cover him; the willows of the brook surround him".

"Under the lotus plant it lies, hidden among the reeds in the math. The lotuses conceal it in their shadow; the poplars by the stream surround it".

Wait... Just read this passage...

Psalm 19

For the director of music. A psalm of David.

1
The heavens declare the glory of God;
  the skies proclaim the work of his hands.
2
Day after day they pour forth speech;
  night after night they reveal knowledge.
3
They have **no speech**, they use **no words**;
  no sound is heard from them.
4
Yet their voice goes out into all the earth,
  their words to the ends of the world.

In the heavens God has pitched a tent for the sun.
5
    It is like a bridegroom coming out of his chamber,
  like a champion rejoicing to run his course.
6
It rises at one end of the heavens
  and makes its circuit to the other;
  nothing is deprived of its warmth.
7
The law of the Lord is perfect,
  refreshing the soul.
The statutes of the Lord are trustworthy,
  making wise the simple.
8
The precepts of the Lord are right,
  giving joy to the heart.
The commands of the Lord are radiant,
  giving light to the eyes.
9
The fear of the Lord is pure,
  enduring forever.
The decrees of the Lord are firm,
  and all of them are righteous.
10
They are more precious than gold,
  than much pure gold;
they are sweeter than honey,
  than honey from the honeycomb.
11
By them your servant is warned;

in keeping them there is great reward.

12

But who can discern their own errors?
   Forgive my hidden faults.

13

Keep your servant also from willful sins;
   may they not rule over me.
Then I will be blameless,
   innocent of great transgression.

14

May these words of my mouth and this meditation
of my heart
   be pleasing in your sight,
   Lord, my Rock and my Redeemer.

That is Radio Waves, just like Job!

Job 38: 34-35

"Can you raise your voice to the clouds
   and cover yourself with a flood of water?
"Do you send the lightning bolts on their way?
   Do they report to you, 'Here we are'?

YOU HAVE TO HAVE ALL 3 OF THESE!!!
PRAYER BOOKS

ALMOND JOY VS 'HARLEM'

The amygdala is an almond-shaped nucleus located deep and medially within the temporal lobe and is thought to play a crucial role in the regulation of emotional processes. GABAergic neurotransmission inhibits the amygdala and prevents us from generating inappropriate emotional and behavioral responses.

Magnesium aka Chlorophyll for the win! The Heart mineral just so happens to be the Amygdala mineral huh? Goto AmricanHealer.Website

B-Vitamins   -   Homosyteine/Cortisol metabolism
Berries or amazon.com lol

I'm gonna pick on Harlem for a second.

Harlem is best example I can think of for this quick powerful lesson. My family is from Harlem, old skool Harlem Royalty. The thing is I learned in school that Harlem was made up! I was in my imagination, there was no such a place as Harlem on the Maps in my school. Just 11 x 13 Miles of Manhattan. What is the time and space of your state? I made it all the way to Clark JHS before figuring this out driving everyone crazy with this... Telling people Harlem was not real, it was in their mind. The muslim brothers said that it was the whiteman's tricknology LMAO...

You know how my mind works by now though, I was all over this... Years later I would learn that every where was fake, then later still, that light and sound only exist in our mind's eye (Heru)!!!

Mind - "**that which feels**, wills, and thinks; the intellect," late 12c., mynd, from Old English gemynd "memory, remembrance; **state of being remembered**; thought, purpose; conscious mind, intellect, intention," Proto-Germanic *ga-mundiz (source also of Gothic muns "**thought**," munan "to think;" Old Norse minni "mind;" German Minne (archaic) "love," originally "memory, loving memory"), from suffixed form of PIE root *men- (1) "to think," with derivatives referring to qualities of mind or states of thought.

The meaning "**mental faculty**, the thinking process" is from c. 1300. The sense of "intention, purpose" is from c. 1300. From late 14c. as "frame of mind. mental disposition," also "way of thinking, opinion."

"**Memory**," one of the oldest senses, now is almost obsolete except in old expressions such as bear in mind (late 14c.), call to mind (early 15c.), keep in mind (late 15c.). The expression time out of mind "indefinite long period of time" is roughly from mid-14c. (tyme of whilk no mynd es), later, in English law, "before Richard I" (1189).

Mind's eye "**mental view or vision**, remembrance" is from early 15c. To pay no mind "disregard" is recorded by 1910, American English dialect. To make up (one's) mind "determine, come to a definite conclusion" is by 1784. To have a mind "be inclined or disposed" (to do something) is by 1540s; to have half a mind to "to have one's mind half made up to (do something)" is recorded from 1726. Out of (one's) mind "mad, insane" is from late 14c.; out of mind "forgotten" is from c. 1300; phrase time out of mind "time beyond people's memory" is attested from early 15c.

My head is hands and feet. I feel all my best faculties concentrated in it. My instinct tells me that my head is an organ for burrowing, as some creatures use their snout and fore-paws, and with it I would mine and burrow my way through these hills. I think that the richest vein is somewhere hereabouts; so by the divining rod and thin rising vapors I judge; and here I will begin to mine. [Thoreau, "Walden"]

mid-14c., "to remember, call to mind, take care to remember," also "to remind oneself," from mind (n.). The Old English verb was myngian, myndgian, from West Germanic *munigon "to remind." Meaning "**perceive, notice**" is from late 15c.; that of "to give heed to, pay attention to" is from 1550s; that of "be careful about" is from 1737. Sense of "object to, dislike" is from c. 1600. Meaning "to take care of, look after" is from 1690s. Related: Minded; minding.
Negative use "(not) to care for, to trouble oneself with" is attested from c. 1600; never mind "don't let it trouble you" is by 1778; the meiotic expression don't mind if I do is attested from 1847.

State - [mode or form of existence] c. 1200, stat, "**circumstances**, position in society, temporary attributes of a person or thing, conditions," from Old French estat "position, condition; status, stature, station," and directly from Latin status "a station, position, **place**; way of standing, posture; order, arrangement, condition," figuratively "standing, rank; public order, community organization."

This is a noun of action from the past-participle stem of stare "**to stand**" (from PIE root *sta- "to stand, make or be firm"). Some Middle English senses are via Old French estat (French état; see estate). The Latin word was adopted into other modern Germanic languages (German, Dutch staat) but chiefly in the political senses only. The meanings "**physical condition as regards form or structure**," "particular condition or phase," and "condition with reference to a norm" are attested from c. 1300. The meaning "**mental or emotional condition**" is attested from 1530s (the phrase state of mind is attested by 1749); the specific colloquial sense of "an agitated or perturbed condition" is from 1837.

The meaning "splendor of ceremony, etc., appropriate to high office; dignity and pomp befitting a person of high degree" is from early 14c. Hence to lie in state "be ceremoniously exposed to view before interment" (1705) and keep state "conduct oneself with pompous dignity" (1590s).

He [the President] shall from time to time give to the Congress Information of the State of the Union, and recommend to their Consideration such Measures as he shall judge necessary and expedient. [U.S. Constitution, Article II, Section iii]

1590s, "**to set in a position**, fix (a date, etc.)," from state (n.1) "circumstances, position." The sense of "<u>**declare**</u>, <u>**recite**</u>, <u>**set down in detail in words**</u>" is attested by 1640s from the notion of "placing" the words on the record. Related: Stated; stating.

"political organization of a country; supreme civil power, the government; the whole people considered as a body politic," 1530s, **from special use of state** (n.1); this sense grew out of the meaning "condition of a country" with regard to government, prosperity, etc. (late 13c.), from Latin phrases such as status rei publicæ "condition (or existence) of **<u>the republic</u>**."

The sense of "a semi-independent political entity under a federal authority, one of the bodies politic which together make up a federal republic" is from 1774. The British North American colonies occasionally were called states as far back as 1630s.

State rights in U.S. political sense is attested from 1798 (the form states rights is recorded by 1824): the doctrine that states retain all rights and privileges not delegated to the federal government in the Constitution, in its extreme form including the power and right of sovereignty.

Often contrasted with ecclesiastical power in phrase church and state (1580s).

State socialism attested from 1850 as "a

scheme of government favoring enlargement of state functions as the directest way to achieve socialist goals."

States - late 14c., **states of the realm**, "one of the major classes constituting **the body politic** and participating in parliament;" see state (n.1) and compare estate. It was used 16c.-17c. of the Netherlands and the States has been short for the United States of America since 1777.

Get it? If you jumped in a time machine, and went back 1,000 years, in the very place you are reading this book, where would you be? I am in Michigan right now, 1,000 years ago Michigan did not exist. The land was here, people lived on it, but it wasn't Michigan. How did this place become Michigan? The first thing the Bible should teach is that all alphabets are codes. Michigan is a code like a mental code for coordinates, we agree collectively is Michigan. The squirrels and birds that have been here for thousands of years don't know what Michigan is or where Michigan is. The Geese fly south and then fly back to Michigan every year, what do they call Michigan? Different rulers of the Land and totally different languages. A ruler measures correct?

Ruler - late 14c., "one who rules or governs," especially in reference to superior or sovereign authority, agent noun from rule (v.). The meaning "instrument used for guiding a pen, etc. in making straight lines" is by c. 1400 (compare rule (n.), the older word for this).

Land - Old English lond, land, "ground, soil," also "definite portion of the earth's surface, home region of a person or a people, territory marked by political boundaries," from Proto-Germanic *landja- (source also of Old Norse, Old Frisian Dutch, Gothic land, German Land), perhaps from PIE *lendh- (2) "land, open land, heath" (source also of Old Irish land, Middle Welsh llan "an open space," Welsh llan "enclosure, church," Breton lann "heath," source of French lande; Old Church Slavonic ledina "waste land, heath," Czech lada "fallow land"). But Boutkan finds no IE etymology and suspects a substratum word in Germanic,

Etymological evidence and Gothic use indicates the original Germanic sense was "a definite portion of the earth's surface owned by an individual or home of a nation." The meaning was early extended to "solid surface of the earth," a sense which once had belonged to the ancestor of Modern English earth (n.). Original senses of land in English now tend to go with country. **To take the lay of the land is a nautical expression**. In the American English exclamation land's sakes (1846) **land is a euphemism for Lord**.

What? Land is a euphemism for Lord? In the Bible Lord means _____. Wait so remember

we briefly discuss observation meaning measurement, is it just a co-inky dink that a Ruler is a measuring Stick (rod) and also means the Most High Authority on the Land? Did you know a Ruler is a Measuring Rod or that a Measuring Rod is a Gauge?

Gauge - "**ascertain by exact measurements**," mid-15c., from Anglo-French gauge (mid-14c.), from Old North French gauger "standardize, calibrate, measure" (Old French jaugier), from gauge "**gauging rod**," a word of unknown origin. Perhaps from Frankish *galgo "**rod, pole for measuring**" or another Germanic source (compare Old Norse gelgja "pole, perch," Old High German galgo; see gallows). Related: Gauged; gauging. The figurative use is from 1580s. "**The spelling variants gauge and gage have existed since the first recorded uses in Middle English**, though in American English gage is found exclusively in technical uses" [Barnhart].

early 15c., "**fixed standard of measure**" (surname Gageman is early 14c.), from Old North French gauge "**gauging rod**" (see gauge (v.)). Meaning "**instrument for measuring**" is from 1670s; meaning "distance between rails on a railway" is from 1841.

Railway-gage, **the distance between perpendiculars** on the insides of the heads of the two rails of a track. Standard gage is 4 feet 8 1/2 inches; anything less than this is narrow gage; anything broader is broad gage. The dimension was fixed for the United States by the wheels of the British locomotive imported from the Stephenson Works in 1829. [Century Dictionary]

The Observer of the People... Wait... You wrestle with these jewels...

Language - late 13c., langage "**words**, what is said, conversation, talk," from Old French langage "speech, words, oratory; **a tribe**, **people**, **nation**" (12c.), from Vulgar Latin *linguaticum, from Latin lingua "**tongue**," also "speech, language" (from PIE root *dnghu- "tongue"). The -u- is an Anglo-French insertion (see gu-); it was not originally pronounced.

The meaning "manner of expression" (vulgar language, etc.) is from c. 1300. The meaning "a language" (as English, French, Arabic, etc.) is from c. 1300; Century Dictionary (1897) defines this as: "**The whole body of uttered signs employed and understood by a given community as expressions of its thoughts**; the aggregate of words, and of methods of their combination into sentences, used in a community for communication and record and for carrying on the processes of thought." Boutkan (2005) writes: "In general, language unity exists as long as the language is capable of carrying out common innovations, but this does not preclude profound differences among dialects."

In Middle English the word also was used of dialects:

Mercii, þat beeþ men of myddel Engelond[,] vnderstondeþ bettre þe side langages, norþerne and souþerne, þan norþerne and souþerne vnderstondeþ eiþer oþer. [John of Trevisa, translation (late 14c.) of Bartholomew Glanville's "De proprietatibus rerum"]

In oþir inglis was it drawin, And turnid ic haue it til ur awin Language of the norþin lede, Þat can na noþir inglis rede. ["Cursor Mundi," early 14c.]

Wait did you not know that Language was just the words but the actual people, and the actual "Nation"?

The place is the words... Words are moor powerful than you thoth huh? Lets keep going, prayers are made of words right? Marinate on that for a minute, did God not create all matter with the word and the word was made Flesh, no? Wakey, wakey....

People work like CellPhones...cell phones and cordless phones use radiofrequency (RF) radiation to send signals. RF radiation is different from other types of radiation. We cooking now.. This book is written very scatter brained for a reason, fyi!

1st we have to accept that our voices can be converted into electromagnetic waves and sent via copper wire.. that's a telephone..

What's different about a cellphone and a telephone? Cellphones can send and receive calls without wire connections of any kind. How does it do this? By using electromagnetic radio waves to send and receive the sounds that would normally travel down wires.

Phone - **word-forming element** meaning "voice, sound," also "speaker of," from Greek phōnē "voice, sound" of a human or animal, also "tone, voice, pronunciation, speech," from PIE root *bha- (2) "to speak, say, tell" (source also of Latin for, fari "to speak," fama "talk, report").
"**elementary sound of a spoken language, one of the primary elements of utterance**," 1866, from Greek phōnē "sound, voice" (from PIE root *bha- (2) "to speak, tell, say").

by 1878 [Des Moines Register, May 16], colloquial shortening of telephone (n.), "generally applied to the receiver, but sometimes to the whole apparatus" [Century Dictionary, 1895]. Phone book "publication listing telephone numbers and their associated names" is by 1920; phone booth "small enclosure or stall provided with a public pay-telephone" is by 1906; phone bill "statement of charges for telephone service" is by

1901; phone number (short for telephone number) is by 1906.

Cell - early 12c., "**small monastery**, subordinate monastery" (from Medieval Latin in this sense), later "**small room for a monk or a nun in a monastic establishment**; **a hermit's dwelling**" (c. 1300), from Latin cella "small room, store room, hut," related to Latin celare "**to hide, conceal**" (from PIE root *kel- (1) "to cover, conceal, save").

From "monastic room" the sense was extended to "prison room" (1722). The word was used in 14c., figuratively, of **brain "compartments" as the abode of some faculty**; it was used in biology by 17c. of various cavities (wood structure, segments of fruit, bee combs), gradually focusing to the modern sense of "**basic structure of all living organisms**" (which OED dates to 1845).

Electric battery sense is from 1828, based on the "compartments" in very early types. The meaning "**small group of people working within a larger organization**" is from 1925. Cell-body is from 1851, cell-division from 1846, cell-membrane from 1837 (cellular membrane is by 1732), cell wall is attested from 1842.

Cellular - 1753, "**consisting of or resembling cells**," with reference to tissue, from Modern Latin cellularis "of little cells," from cellula "little cell," diminutive of cella "**small

room" (see cell). <u>Of mobile phone systems</u> (**in which the area served is divided into "cells" of a few square miles served by transmitters**), 1977. Related: Cellularity.

Mantra - A mantra (Pali: mantra) or mantram (Devanagari: मन्त्रम्) is **a sacred utterance**, a numinous sound, a syllable, word or **phonemes**, or group of words (most often in an Indo-Iranian language like Sanskrit or Avestan) **believed by practitioners to have religious, magical or spiritual powers**. Some mantras have a syntactic structure and a literal meaning, while others do not.

Phoneme - "**distinctive sound or group of sounds**," 1889, from French phonème, from Greek phōnēma "a sound made, voice," from phōnein "to sound or speak," from phōnē "sound, voice" (from PIE root *bha- (2) "to speak, tell, say"). Related: Phonematic.

You think you know why Tesla was 'killed breathing'...

Genesis 11

And the whole earth was of one language, and of one speech.
2 And it came to pass, as they journeyed from the east, that they found a plain in the land of Shinar;

and they dwelt there.

3 And they said one to another, Go to, let us make brick, and burn them thoroughly. And they had brick for stone, and slime had they for morter.

4 And they said, Go to, let us build us a city and a tower, whose top may reach unto heaven; and let us make us a name, lest we be scattered abroad upon the face of the whole earth.

5 And the Lord came down to see the city and the tower, which the children of men builded.

6 And the Lord said, Behold, the people is one, and they have all one language; and this they begin to do: and now nothing will be restrained from them, which they have imagined to do.

7 Go to, let us go down, and there confound their language, that they may not understand one another's speech.

8 So the Lord scattered them abroad from thence upon the face of all the earth: and they left off to build the city.

9 Therefore is the name of it called Babel; because the Lord did there confound the language of all the earth: and from thence did the Lord scatter them abroad upon the face of all the earth.

10 These are the generations of Shem: Shem was an hundred years old, and begat Arphaxad two years after the flood:

11 And Shem lived after he begat Arphaxad five hundred years, and begat sons and daughters.

12 And Arphaxad lived five and thirty years, and begat Salah:

13 And Arphaxad lived after he begat Salah four hundred and three years, and begat sons and daughters.

14 And Salah lived thirty years, and begat Eber:

15 And Salah lived after he begat Eber four hundred and three years, and begat sons and daughters.

16 And Eber lived four and thirty years, and begat Peleg:

17 And Eber lived after he begat Peleg four hundred and thirty years, and begat sons and daughters.

18 And Peleg lived thirty years, and begat Reu:

19 And Peleg lived after he begat Reu two hundred and nine years, and begat sons and daughters.

20 And Reu lived two and thirty years, and begat Serug:

21 And Reu lived after he begat Serug two hundred and seven years, and begat sons and daughters.

22 And Serug lived thirty years, and begat Nahor:

23 And Serug lived after he begat Nahor two hundred years, and begat sons and daughters.

24 And Nahor lived nine and twenty years, and begat Terah:

25 And Nahor lived after he begat Terah an hundred and nineteen years, and begat sons and daughters.

26 And Terah lived seventy years, and begat Abram, Nahor, and Haran.

27 Now these are the generations of Terah: Terah begat Abram, Nahor, and Haran; and Haran begat

Lot.

28 And Haran died before his father Terah in the land of his nativity, in Ur of the Chaldees.

29 And Abram and Nahor took them wives: the name of Abram's wife was Sarai; and the name of Nahor's wife, Milcah, the daughter of Haran, the father of Milcah, and the father of Iscah.

30 But Sarai was barren; she had no child.

31 And Terah took Abram his son, and Lot the son of Haran his son's son, and Sarai his daughter in law, his son Abram's wife; and they went forth with them from Ur of the Chaldees, to go into the land of Canaan; and they came unto Haran, and dwelt there.

32 And the days of Terah were two hundred and five years: and Terah died in Haran.

Languages are the people and nations remember? Add that to the fact that sound and light only exist in your mind.

Tesla was going to fire up that Radio antenna, which NASA - National Aeronautics and Space AdministrationESA - European Space AgencyCNSA - China National Space AdministrationRoscosmos - Russian Federal Space AgencyJAXA - Japan Aerospace Exploration AgencyISRO - Indian Space Research OrganisationCSA - Canada Space Agency, are all really hiding this truth! Yall traded God for Anal Chakra Activations don't get mad with me... In 1901 construction of Tesla's new laboratory at

Wardenclyffe on Long Island began. Tesla's Long Island laboratory included a 60-meter-tall tower, known as the Wardenclyffe Tower. This imposing antenna was specially designed and constructed for the wireless transmission of energy. There was no space agencies during this time, it was Tesla, Max Planck and Einstein collectively that pushed the science at least in public, behind the scenes it was the Bible & Satanists.

Tesla's inventions included:

- AC Power (alternating current)
- Tesla Coil
- Magnifying Transmitter
- Tesla Turbine
- Shadowgraph
- Radio
- Neon Lamp
- Hydroelectric Power
- Induction Motor
- Radio Controlled Boat

Sidebar: Ham Radio Just another instance of Framing Ham... The word "HAM" as applied to 1908 was the station CALL of the first amateur wireless station operated by some amateurs of the Harvard Radio Club. They were Albert S. Hyman, Bob Almy, and Poogie Murray. At first they called their station "HYMAN-ALMY-MURRAY".

1st we have to accept that our voices can be converted into electromagnetic waves and sent via copper wire.. that's a telephone..

2nd Connect your feet n hands together, or hands together while screwing in (grounding), or being in water... or holding hands with others, or during sex (especially heart to heart with hands n feet palms closed)... Boost your signal!

Who are the most persecuted people on Earth? Black People & Christians, clearly being Black & Christian is asking for it! Lets examine that a bit. We all know that people born with height are better on average at basketball, people that are wide are better at football, people with long attention spans are better readers etc... If we start saying there may be people better at prayer or communicating with God, you need to go into witness protection!

The facts are the deeper the concentration of EuMelanin, the more NeuroMelanin especially in the spine and heart, the higher your potential to convert radio waves into "Divine Inspiration".

Investigation of the spinal cord as a natural receptor antenna for incident electromagnetic waves and possible impact on the central nervous system

Sevaiyan Balaguru 1, Rajan Uppal, Ravinder Pal Vaid,

CHASE DUQUESNAY

Balasubramaniam Preetham Kumar

Abstract

The effects of electromagnetic field (EMF) exposure on biological systems have been studied for many years, both as a source of medical therapy and also for potential health risks. In particular, the mechanisms of EMF absorption in the human or animal body is of medical/ engineering interest, and modern modelling techniques, such as the Finite Difference Time Domain (FDTD), can be utilized to simulate the voltages and currents induced in different parts of the body. The simulation of one particular component, the spinal cord, is the focus of this article, and this study is motivated by the fact that the spinal cord can be modelled as a linear conducting structure, capable of generating a significant amount of voltage from incident EMF. In this article, we show, through a FDTD simulation analysis of an incoming electromagnetic field (EMF), that the spinal cord acts as a natural antenna, with frequency dependent induced electric voltage and current distribution. The multi-frequency (100-2400 MHz) simulation results show that peak voltage and current response is observed in the FM radio range around 100 MHz, with significant strength to potentially cause changes in the CNS. This work can contribute to the understanding of the mechanism behind EMF energy leakage into the CNS, and the possible contribution of the latter energy leakage towards the weakening of the blood brain barrier (BBB), whose degradation is associated with the progress of many diseases, including Acquired Immuno-Deficiency Syndrome (AIDS).

This is why 'they' created the Music Industry.

Muse - "to **reflect**, ponder, **meditate**; **to be absorbed in thought**," mid-14c., from Old French muser (12c.) "to ponder, dream, wonder; loiter, waste time," which is of uncertain origin; the explanation in Diez and Skeat is literally "to stand with one's nose in the air" (or, possibly, "to sniff about" like a dog who has lost the scent), from muse "muzzle," from Gallo-Roman *musa "snout," itself a word of unknown origin. The modern word probably has been influenced in sense by muse (n.). Related: Mused; musing.

late 14c., "one of the nine Muses of classical mythology," daughters of Zeus and Mnemosyne, protectors of the arts; from Old French Muse and directly from Latin **Musa**, from Greek **Mousa**, "the Muse," also "**music, song**," ultimately from PIE root *men- (1) "to think." Meaning "**inspiring goddess of a particular poet**" (with a lower-case m-) is from late 14c.

The traditional names and specialties of the nine Muses are: Calliope (epic poetry), Clio (history), Erato (love poetry, lyric art), Euterpe (music, especially flute), Melpomene (tragedy), Polymnia (hymns), Terpsichore (dance), Thalia (comedy), Urania (astronomy).

ic - Middle English -ik, -ick, word-forming element making adjectives, "**having to do with**, **having the nature of**, being, made of, caused by, **similar to**," from French -ique and directly from Latin -icus or from cognate Greek -ikos "in the manner of; pertaining to." From PIE adjective suffix *-(i)ko, which also yielded Slavic -isku, adjectival suffix indicating origin, the source of the -sky (Russian -skii) in many surnames. In chemistry, indicating a higher valence than names in -ous (first in benzoic, 1791).

In Middle English and after often spelled -ick, -ike, -ique. Variant forms in -ick (critick, ethick) were common in early Modern English and survived in English dictionaries into early 19c. This spelling was supported by Johnson but opposed by Webster, who prevailed.

Amuse - late 15c., "**to divert the attention**, **beguile**, delude," from Old French amuser "fool, tease, hoax, entrap; **make fun of**," literally "cause to muse" (as a distraction), from a "at, to" (from Latin ad, but here probably a causal prefix) + muser "ponder, stare fixedly" (see muse (v.)).

The original English senses are obsolete; the meaning "divert from serious business, tickle the fancy of" is recorded from 1630s, but through 18c. the primary meaning was "deceive, cheat" by first occupying the attention. "The word was not in reg. use bef. 1600, and was not used by Shakespere" [OED]. Bemuse retains more of the original meaning. Greek amousos meant "**without Muses**," hence "**uneducated**."

We are the most ignorant, diseased, high, crazy, criminal people on earth, so why does everyone follow us? We make up %13 of America and almost half of that is in prison or lost their voting rights, why do we control elections? We are the voices that the world follows, NO MATTER WHAT! Why? We are tapped into the SkyFi!!! The only way to control what goes on here is to control us, control the radio waves, poison our bodies etc... God is your Super Power, It was not until Satanism took over, the opioid crisis is not a crisis! Oh trust there is a small group that thrives off this! Not the money, the zombie control. My little brother just died addicted to fentanyl and heroine.

God is Music

Nature is his Producer

Pigments are his Beat Machine

The code word for tapping into this system is FREE ENERGY. Never forget energy = information. Free Energy equals Free Information, its in the Air. Just start prayer, Christian, Jewish, Mulsim, Buddhist,

get going...
YOU HAVE TO HAVE ALL 3 OF THESE!!!
PRAYER BOOKS

  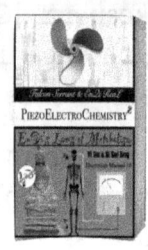

REPS...

You repeat a code to give it strength.

Calisthenics is a sacred form of prayer and meditation ... if done to the right music, in the right **setting**.

Setting - c. 1300, "**a fixed time for meals**;" late 14c., "fact or action of putting (something) in a place or position; a placing, a planting," also "a place, location, site; the manner or position in which anything is fixed," verbal noun from set (v.).

The surgical sense, with reference to broken bones, etc., is from early 15c. (Chauliac). In reference to the sinking of the sun, moon, or other heavenly bodies below the horizon, from c. 1400. Also in Middle English "**act of creation**; **thing created**" (c. 1400). In reference to mounts for jewels, etc. from 1815; the meaning "background, history, environment" is attested from 1841. By 1871 as "act, result, or process of fitting to music." The theatrical sense of "the mounting of a play or opera for the stage" is by 1841. The meaning "set of cutlery, crockery, or both for a single place at table is by 1952.

Set - "**act of setting; state or condition of being set**" (originally of the sun or another heavenly body), mid-14c., from set (v.) or its identical past participle. Old English had set "seat," in plural "camp; stable," but OED finds it "doubtful whether this survived beyond OE." Compare set (n.1).

Disparate senses collect under this word because of the many meanings given the verb. The sense of "manner or position in which something is set" is by 1530s, hence "general movement, direction, drift, tendency, inclination" (of mind, character, policy, etc.), by 1560s.

The meaning "**permanent change of shape caused by pressure**; a bend, warp, **kink**" is by 1812; that of "action of hardening," by 1837. Hence "action or result of fixing the hair when damp so that it holds the desired style" (1933).

"Something that has been set" (1510s), hence the use in tennis, "set of six games which counts as a unit" (1570s) and set-point "state of the game at which one side or player needs only one point to win the set" (by 1928).

The theatrical meaning "scenery for an individual scene in a play, etc.," is by 1859, from the past-participle adjective. It later was extended in movie and television production to the place or area where filming takes place.

Set (n.1) and set (n.2) are not always distinguished in dictionaries; OED has them as two entries, Century Dictionary as one. The difference of opinion seems to be whether the set meaning "group, grouping" (here (n.2)) is a borrowing of the unrelated French word that sounds like the native English one, or a borrowing of the sense only, which was absorbed into the English word.

Don't fret bodybuilders... Steel is a conductor and an excellent choice for conducting both electricity and heat. Its metallic properties, such as the presence of free electrons, enable it to facilitate the flow of electric current with ease. Additionally, steel's ability to conduct heat allows for a more efficient transfer of thermal energy. Pumping iron is equally amazing as a meditation... High calcium bad, high magnesium good! Low magnesium (greens/chlorophyll) causes depression/loneliness!

"Dietary magnesium restriction reduces amygdala–hypothalamic GluN1 receptor complex levels in mice": Erratum.

Ghafari, M., Whittle, N., Miklósi, A. G., Kotlowski, C., Schmuckermair, C., Berger, J., Bennett, K. L., Singewald, N., & Lubec, G. (2015).

Abstract

Reports an error in "Dietary Magnesium

Restriction Reduces Amygdala–Hypothalamic GluN1 Receptor Complex Levels in Mice" by M. Ghafari, N. Whittle, A. G. Miklósi, C .Kotlowsky, C. Schmuckermair, J. Berger, K. L. Bennett, N. Singewald and G. Lubec (Brain Structure & Function, Advanced Online Publication, May 8, 2014, np). The last name of Caroline Kotlowski, the fourth author of this paper, was inadvertently misspelled when this article was originally published. The correct spelling is "Kotlowski". (The following abstract of the original article appeared in record 2014-19623-001). Reduced daily intake of magnesium (Mg2+) is suggested to contribute to depression. Indeed, preclinical studies show dietary magnesium restriction (MgR) elicits enhanced depression-like behaviour establishing a causal relationship. Amongst other mechanisms, Mg2+ gates the activity of N-methyl-D-asparte (NMDA) receptors; however, it is not known whether reduced dietary Mg2+ intake can indeed affect brain NMDA receptor complexes. Thus, the aim of the current study was to reveal whether MgR induces changes in brain NMDA receptor subunit composition that would indicate altered NMDA receptor regulation. The results revealed that enhanced depression-like behaviour elicited by MgR was associated with reduced amygdala–hypothalamic protein levels of GluN1-containing NMDA complexes. No change in GluN1 mRNA levels was observed indicating posttranslational changes were induced by dietary Mg2+ restriction.

To reveal possible protein interaction partners, GluN1 immunoprecipitation and proximity ligation assays were carried out revealing the expected GluN1 subunit association with GluN2A, GluN2B, but also novel interactions with GluA1, GluA2 in addition to known downstream signalling proteins. Chronic paroxetine treatment in MgR mice normalized enhanced depression-like behaviour, but did not alter protein levels of GluN1-containing NMDA receptors, indicating targets downstream of the NMDA receptor. Collectively, present data demonstrate that dietary MgR alters brain levels of GluN1-containing NMDA receptor complexes, containing GluN2A, GluN2B, AMPA receptors GluA1, GluA2 and several protein kinases. These data indicate that the modulation of dietary Mg2+ intake may alter the function and signalling of this receptor complex indicating its involvement in the enhanced depression-like behaviour elicited by MgR.

Fyi we have just the thing for you on AmericanHealer.website

Cbd/thc are great too... cbd is better tho.... Thc over stimulates then suppresses the sympathetic nerves then when you quit or just sober its hyperactive, leading to fake anxiety, its real but there is no external cause!

CHASE DUQUESNAY

YOU HAVE TO HAVE ALL 3 OF THESE!!! PRAYER BOOKS

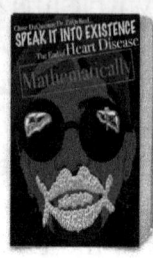

ANOTHER DEEP DIVE?

okok

Lets go in in…

99% of religious scholars fail because they fall into this categories:

Eisegesis is the interpretation of a text (as of the Bible) by reading into it one's own ideas, as opposed to Exegesis. In fact every pro or anti bible scholar I have ever disagreed with on critical terms says the same thing "you can't use the word of God in a literal sense".

Literal - (of a translation) representing the exact words of the original text.

Smh… This word salad culinary artist routine is what they use to bend the words and run people away. A million different churches and a million different interpretations. That leads the common sense logical person to say to themselves, these words CLEARLY don't mean anything. People just bend them to serve their own ends.

Do this and tell them who you teacher is! Don't say God taught EnQi, EnQi taught me, so its not a lie if I say God taught me... It is, we have to have a global re-ordering! Order.

Question 1 Did God create the world with the word? Yes
Question 2 Was the word made Flesh? Yes

Question 3 Is all flesh the same? No

Question 4 Is there a type of Flesh more conducive to sound and light? Yes

Question 5 What type of Flesh is that? _____ _____ Flesh.

Question 6 Is God Light? Yes

Question 7 Can all Flesh absorb ElectroMagnetic **Waves** Equally? No

Question 8 Can all Flesh absorb Mechanical **Waves** Equally? No

Question 9 Does God have a language? Yes

Question 10 What is that Language? Song & Dance

Question 11 Is there a group of people Naturally more talented at song and dance? Yes

Question 12 Are the people naturally more talented at song and dance, also in possession of the flesh most conducive to absorbing light & sound?

Question 13 Are the people naturally more talented at song and dance, that are also in possession of the flesh most conducive to absorbing light & sound, also the most hated and persecuted?

Question 14 Is a Covenant between God & Man Electromagnetism or Light? Yes

Question 15 How does man perceive Light? Pigment.

Question 16 Did God create Flesh & Pigment? Yes

Question 17 Why doesn't everyone have the same Flesh?... Ill stop.

Let's lighten the mood (pun intended). We just have to be on the look out for Charles Darwin's Theory of Evolution and it's religious twin Supersessionism.

Supersessionism - Supersessionism, also called replacement theology, is the Christian doctrine that the Christian Church has superseded the Jewish people, assuming their role as God's covenanted people, thus asserting that the New Covenant through Jesus Christ has superseded or replaced the Mosaic covenant.

Matthew 5:17-20 reads as follows: 17 "**Do not think that I came to abolish the Law or the**

Prophets; I did not come to abolish but to fulfill. 18 "For truly I say to you, until heaven and earth pass away, <u>not the smallest letter or stroke shall pass from the Law until all is accomplished</u>.

The Bible says exactly what I am saying, I just have science that agrees with the Bible as it should.

Isaiah 49 1-6

Listen, O isles, unto me; and hearken, ye people, from far; The Lord hath called me from the womb; from the bowels of my mother hath he made mention of my name.
2 And he hath made my mouth like a sharp sword; in the shadow of his hand hath he hid me, and made me a polished shaft; in his quiver hath he hid me;
3 And said unto me, Thou art my servant, O Israel, in whom I will be glorified.
4 Then I said, I have laboured in vain, I have spent my strength for nought, and in vain: yet surely my judgment is with the Lord, and my work with my God.
5 And now, saith the Lord that formed me from the womb to be his servant, to bring Jacob again to him, Though Israel be not gathered, yet shall I be glorious in the eyes of the Lord, and my God shall be my strength.
6 And he said, **It is a light thing** that thou shouldest be my servant to raise up the tribes of Jacob, and to restore the preserved of Israel: **I will**

also give thee for a light to the Gentiles, that thou mayest be my salvation unto the end of the earth.

The Gentiles have to follow God's Children to get the "light".

Lets go to the Flood story... We will compare the Bible version to the Sumerian version... There is a missing ting deh....

The mystery, the missing section of the resolution... I dreamed it as a kid... I didn't know what I was having nightmares of....but it was floods... after my heart stopped at 24 year & 4 months old.... It began coming back to me... leading me here!

Before the missing section, the gods have decided to send a flood to destroy humanity. Enki, god of the underworld sea of fresh water and equivalent of Babylonian Ea, warns Ziusudra, the ruler of Shuruppak, to build a large boat, though the directions for the boat are also lost.

When the tablet resumes, it describes the flood. A terrible storm rages for seven days and nights. "The huge boat had been tossed about on the great waters." Then Utu (Sun) appears and Ziusudra opens a window, prostrates himself, and sacrifices an ox and a sheep.

After another break, the text resumes with the

flood apparently over, and Ziusudra prostrating himself before An (Sky) and Enlil (Lordbreath), who give him "breath eternal" for "preserving the animals and the seed of mankind". The remainder is lost.

The Epic of Ziusudra adds an element at lines 258–261 not found in other versions, that after the river flood "king Ziusudra ... they caused to dwell in the land of the country of Dilmun, the place where the sun rises". In this version of the story, Ziusudra's boat floats down the Euphrates river into the Persian Gulf (rather than up onto a mountain, or up-stream to Kish). The Sumerian word *KUR* in line 140 of the Gilgamesh flood myth was interpreted to mean "mountain" in Akkadian, although in Sumerian, KUR means "mountain" but also "land", especially a foreign country, as well as "the Underworld".

The resolution of character, the devil deal. You create or raise hell when you are angry/violent aka disagreeable! You have a self destruct button, so the entire planet doesn't need to be wiped away, unless we all turn away from god... **The self destruct button is the Almond...** The Satan Clause! If you follow Satan....

Noah and the Flood

9 This is the account of Noah and his family.

Noah was a righteous man, blameless among the

people of his time, and he walked faithfully with God. **10** Noah had three sons: Shem, Ham and Japheth.

11 Now the earth was **corrupt in God's sight** and was full of **violence**. **12** God saw ho**w corrupt the earth had become, f**or all the people on earth had corrupted their ways. **13** So God said to Noah, "I am going to put an end to all people, for the earth is **filled with violence** because of them. I am surely going to destroy both them and the earth. **14** So make yourself an ark of cypress[a] wood; make rooms in it and coat it with pitch inside and out. **15** This is how you are to build it: The ark is to be three hundred cubits long, fifty cubits wide and thirty cubits high. **16** Make a roof for it, leaving below the roof an opening one cubit[c] high all around. Put a door in the side of the ark and make lower, middle and upper decks. **17** I am going to bring floodwaters on the earth to destroy all life under the heavens, every creature that has the breath of life in it. Everything on earth will perish. **18** But I will establish my covenant with you, and you will enter the ark—you and your sons and your wife and your sons 'wives with you. **19** You are to bring into the ark two of all living creatures, male and female, to keep them alive with you. **20** Two of every kind of bird, of every kind of animal and of every kind of creature that moves along the ground will come to you to be kept alive. **21** You are to take every kind of food that is to be eaten and store it away as food for you and for them."

22 Noah did everything just as God commanded him.

Lines 1-203, Tablet XI (note: with supplemental sub-titles and line numbers added for clarity)

Ea leaks the secret plan

1. Utnapishtim tells Gilgamesh a secret story that begins in the old city of Shuruppak on the banks of the Euphrates River.
2. The "great gods" Anu, Enlil, Ninurta, Ennugi, and Ea were sworn to secrecy about their plan to cause the flood.
3. But the god Ea (Sumerian god Enki) repeated the plan to Utnapishtim through a reed wall in a reed house.
4. Ea commanded Utnapishtim to demolish his house and build a boat, regardless of the cost, to keep living beings alive.
5. The boat must have equal dimensions with corresponding width and length and be covered over like Apsu boats.
6. Utnapishtim promised to do what Ea commanded.
7. He asked Ea what he should say to the city elders and the population.
8. Ea tells him to say that Enlil has rejected him and he can no longer reside in the city or set

foot in Enlil's territory.

9. He should also say that he will go down to the Apsu "to live with my lord Ea".

Note: 'Apsu' can refer to a freshwater marsh near the temple of Ea/Enki at the city of Eridu.

Building and launching the boat

1. Carpenters, reed workers, and other people assembled one morning.
2. [missing lines]
3. Five days later, Utnapishtim laid out the exterior walls of the boat **of 120 cubits.**
4. **The sides of the superstructure had equal lengths of 120 cubits.** He also made a drawing of the interior structure.
5. The boat had six decks [?] divided into seven and nine compartments.
6. Water plugs were driven into the middle part.
7. Punting poles and other necessary things were laid in.
8. Three times **3,600 units of raw bitumen were melted in a kiln and three times 3,600 units of oil were used in addition to two times 3,600** units of oil that were stored in the boat.
9. Oxen and sheep were slaughtered and ale, beer, oil, and wine were distributed to the workmen, like at a new year's festival.
10. When the boat was finished, the launch was very difficult. A runway of poles

was used to slide the boat into the water.

11.   Two-thirds of the boat was in the water.

12.   Utnapishtim loaded his silver and gold into the boat.

13.   He loaded "all the living beings that I had."

14.   His relatives and craftsmen, and "all the beasts and animals of the field" boarded the boat.

15.   The time arrived, as stated by the god Shamash, to seal the entry door.

The storm

1.   Early in the morning at dawn a black cloud arose from the horizon.

2.   The weather was frightful.

3.   Utnapishtim boarded the boat and entrusted the boat and its contents to his boat master Puzurammurri who sealed the entry.

4.   The thunder god Adad rumbled in the cloud and storm gods Shullat and Hanish went over mountains and land.

5.   Erragal pulled out the mooring poles and the dikes overflowed.

6.   The Anunnaki gods lit up the land with their lightning.

7.   There was stunned shock at Adad's deeds which turned everything to blackness. The land was shattered like a pot.

8.   All day long the south wind blew rapidly

and the water overwhelmed the people like an attack.

9.   No one could see his fellows. They could not recognize each other in the torrent.

10.   The gods were frightened by the flood and retreated up to the Anu heaven. They cowered like dogs lying by the outer wall.

11.   Ishtar shrieked like a woman in childbirth.

12.   The Mistress of the gods wailed that the old days had turned to clay because "I said evil things in the Assembly of the Gods, ordering a catastrophe to destroy my people who fill the sea like fish."

13.   The other gods were weeping with her and sat sobbing with grief, their lips burning, parched with thirst.

14.   The flood and wind lasted **six days and six nights,** flattening the land.

15.   On the seventh day, the storm was pounding [intermittently?] like a woman in labour.

Calm after the storm

1.   The sea calmed and the whirlwind and flood stopped. All-day long there was quiet. All humans had turned to clay.

2.   The terrain was as flat as a rooftop. Utnapishtim opened a window and felt fresh air on his face.

3.   He fell to his knees and sat weeping,

tears streaming down his face. He looked for coastlines on the horizon and saw a region of land.

4.   The boat lodged firmly on mount Nimush which held the boat for several days, allowing no swaying.

5.   On the **seventh day he released a dove** that flew away but came back to him. He released a swallow, but it also came back to him.

6.   He released a raven that was able to eat and scratch, and did not circle back to the boat.

7.   He then sent his livestock out in various directions.

The sacrifice

1.   He sacrificed a sheep and offered incense at a mountainous ziggurat where **he placed 14 sacrificial vessels** and poured reeds, cedar, and myrtle into the fire.

2.   The gods smelled the sweet odour of the sacrificial animal and gathered like flies over the sacrifice.

3.   Then the great goddess arrived, lifted up her flies (beads), and said

4.   "Ye gods, as surely as I shall not forget this lapis lazuli [amulet] around my neck, I shall be mindful of these days and never forget them! The gods may come to the sacrificial offering. But Enlil may not come, because he brought about the flood and annihilated my people without considering [the consequences]."

5. When Enlil arrived, he saw the boat and became furious at the Igigi gods. He said "Where did a living being escape? No man was to survive the annihilation!"

6. Ninurta spoke to Enlil saying "Who else but Ea could do such a thing? It is Ea who knew all of our plans."

7. Ea spoke to Enlil saying "It was you, the Sage of the Gods. How could *you* bring about a flood without consideration?"

8. Ea then accuses Enlil of sending a disproportionate punishment and reminds him of the need for compassion.

9. Ea denies leaking the god's secret plan to Atrahasis (= Utnapishtim), admitting only sending him a dream and deflecting Enlil's attention to the flood hero.

The flood hero and his wife are granted immortality and transported far away[edit]

1. Enlil then boards a boat and grasping Utnapishtim's hand, helps him and his wife aboard where they kneel. Standing between Utnapishtim and his wife, he touches their foreheads and blesses them. "Formerly Utnapishtim was a human being, but now he and his wife have become gods like us. Let Utnapishtim reside far away, at the mouth of the rivers."

2. Utnapishtim and his wife are transported and settled at the "mouth of the rivers".

Pay close attention In the oldest version of the story, Enlil decides to exterminate humanity because **their noise disturbs his sleep**.

There is a order on Earth that must be maintained, God eliminates whatever threatens that order, because it threatens the purpose. Anything that disrupts the Radio...

God is Music

YOU HAVE TO HAVE ALL 3 OF THESE!!!
PRAYER BOOKS

DALA OR DALER, DOLLAR

Does money grow on trees?

Dal - sort of vetch cultivated in the East Indies, 1690s, from Hindi dal "**split pulse**," from Sanskrit dala, from dal "to split."

Dollar - "**monetary unit** or standard of value in the U.S. and Canada," 1550s, **daler**, originally in English the name of a large, silver coin of varying value in the German states, from Low German daler, from German taler (1530s, later thaler), abbreviation of Joachimstaler, literally "(gulden) of Joachimstal," coin minted 1519 from silver from mine opened 1516 near Sankt Joachimsthal, town in Erzgebirge Mountains in northwest Bohemia. **German Tal is cognate with English dale**. The spelling had been modified to dollar by 1600.

The thaler was from 17c. the more-or-less

standardized coin of northern Germany (as opposed to the southern gulden). It also served as a currency unit in Denmark and Sweden (and later was a unit of the German monetary union of 1857-73 equal to three marks).

English colonists in America used the word dollar from 1580s in reference to Spanish peso or "**piece of eight**," also a large silver coin of about the same fineness as the thaler. Due to extensive trade with the Spanish Indies and the proximity of Spanish colonies along the Gulf Coast, the Spanish dollar probably was the coin most familiar in the American colonies and the closest thing to a standard in all of them.

When the Revolution came, it had the added advantage of not being British. It was used in the government's records of public debt and expenditures, and the Continental Congress in 1786 adopted dollar as a unit when it set up the modern U.S. currency system, which was based on the suggestion of Gouverneur Morris (1782) as modified by Thomas Jefferson. None were circulated until 1794.

When William M. Evarts was Secretary of State he accompanied Lord Coleridge on an excursion to Mount Vernon. Coleridge remarked that he had heard it said that Washington, standing on the lawn, could throw a dollar clear across the Potomac. Mr. Evarts explained that a dollar would go further in those days than now. [Walsh]

Phrase dollars to doughnuts "an assured thing, a certainty" (such that one would bet a dollar against a doughnut on it) is attested by 1884; dollar diplomacy "financial imperialism, foreign policy based on financial and commercial interests" is from 1910.

The dollar sign ($) is said to derive from the image of the Pillars of Hercules, stamped with a scroll, on the Spanish **piece of eight**. However, according to the Bureau of Engraving and Printing of the U.S. Department of the Treasury:

[T]he most widely accepted explanation is that the symbol is the result of evolution, independently in different places, of the Mexican or Spanish "P's" for pesos, or piastres, or pieces of eight. The theory, derived from a study of old manuscripts, is that the "S" gradually came to be written over the "P," developing a close equivalent of the "$" mark. It was widely used before the adoption of the United States dollar in 1785.

(Only it Cleary looks like a serpent on a rod, lol.)

Mandala - **symbolic magic circle used by Buddhists in meditation**, 1792, from Sanskrit mandala "**disc, circle**." Adopted 20c. in Jungian psychology as a symbol of unity of the self and completeness.

Dollar was a coin and so was mandala... interesting! Why do we spend money we don't have, to buy things we don't need, to impress and emulate people we don't like or want to be? Fear of exclusion (loneliness).

Amygdala - part of the brain, from Latin amygdalum **"<u>almond</u>"** (which the brain parts resemble), from Greek amygdalē "almond" (see almond). English had also amygdales "the tonsils" (early 15c.), from a secondary sense of the Latin word in Medieval Latin, a translation of Arabic al-lauzatani "the two tonsils," literally "the two almonds," so called by Arabic physicians for fancied resemblance.

Almond - **kernel of the fruit of the almond tree**, c. 1300, from Old French almande, amande, earlier alemondle "almond," from Vulgar Latin ***amendla**, *amandula, from Latin **amygdala** (plural), from Greek **amygdalos** "an almond tree," a word of unknown origin, perhaps from Semitic. Late Old English had amygdales "almonds."
It was altered in Medieval Latin by influence of amandus "loveable." In French it acquired an unetymological -l-, perhaps from Spanish almendra "almond," which got it by influence of the many Spanish words beginning with the Arabic definite article al-. Perhaps through similar confusion, Italian has dropped the first letter entirely (mandorla). As an adjective, applied to eyes shaped like almonds, especially of

certain Asiatic peoples, from 1849.

Do not eat the fruit of the tree of good and evil... huh? When you eat this fruit you surely die!

Genesis 2:17

but of the tree of the knowledge of good and evil, thou shalt not eat of it. For in the day that thou eatest thereof, thou shalt surely die."

Sitting right in plain sight! We are made of the dust of the ground, divine soil. The tree in the midst of the Garden is? We detailed how the brain is actually created from a rod as a embryo.

Numbers 17

And the Lord spake unto Moses, saying,

2 Speak unto the children of Israel, and take of every one of them a rod according to the house of their fathers, of all their princes according to the house of their fathers twelve rods: write thou every man's name upon his rod.

3 And thou shalt write Aaron's name upon the rod of Levi: for one rod shall be for the head of the house of their fathers.

4 And thou shalt lay them up in the tabernacle of the congregation before the testimony, where I will meet with you.

5 And it shall come to pass, that the man's rod, whom I shall choose, shall blossom: and I will make to cease from me the murmurings of the children of Israel, whereby they murmur against you.

6 And Moses spake unto the children of Israel, and every one of their princes gave him a rod apiece, for each prince one, according to their fathers' houses, even twelve rods: and the rod of Aaron was among their rods.

7 And Moses laid up the rods before the Lord in the tabernacle of witness.

8 And it came to pass, that on the morrow Moses went into the tabernacle of witness; and, behold, the rod of Aaron for the house of Levi was budded, and brought forth buds, and bloomed blossoms, and **yielded almonds.**

9 And Moses brought out all the rods from before the Lord unto all the children of Israel: and they looked, and took every man his rod.

10 And the Lord said unto Moses, Bring Aaron's rod again before the testimony, to be kept for a token against the rebels; and thou shalt quite take away their murmurings from me, **that they die not.**

11 And Moses did so: as the Lord commanded him, so did he.

12 And the children of Israel spake unto Moses, saying, Behold, **we die, we perish, we all perish.**

13 Whosoever cometh any thing near unto the tabernacle of the Lord shall die: shall we be **consumed with dying?**

The Rod is associated, even derived from the Egyptian rod called the Was Sceptre which we have identified as the Aorta!!! We showed in the Speak it into Existence book, that the "Almond" causes death through the Was Sceptre, the Amygdala causes Heart Disease through the Aorta!

And the LORD spoke to Moses, saying: "Speak to the children of Israel, and get from them a rod from each father's house, all their leaders according to their fathers 'houses; twelve rods. Write each man's name on his rod. And you shall write Aaron's name on the rod of Levi. For there shall be one rod for the head of each father's house.

a. **Get from them a rod from each father's house**: **A rod** was a symbol of authority because shepherds would use a rod to guide and correct the sheep (Psalm 23:4).

i. Moses, as a shepherd, had **a rod** in his hand when tending sheep in the wilderness (Exodus 4:2). This rod later became known as the rod of God (Exodus

4:20), a symbol of the authority God gave to Moses.

ii. This same **rod** held by Moses demonstrated his authority in action. The rod of God in the hand of Moses:

· Miraculously became a serpent, and then became a rod again (Exodus 7:9-10).

· Turned the waters of the Nile into blood (Exodus 7:17).

· Brought plagues of frogs (Exodus 8:5), lice (Exodus 8:16-17), hail (Exodus 9:23), and locusts (Exodus 10:13).

· Was raised over the Red Sea when it was to be parted (Exodus 14:16).

· Was raised in prayer over Israel in victorious battle (Exodus 17:9).

· Struck the rock and brought water (Numbers 20:11).

iii. The **rod** is also a picture of God's authority over man (Psalm 2:9, 23:4, 89:32; Isaiah 10:24, 11:4, Ezekiel 20:37). Jesus, in His divine authority, was given the title the Rod (Isaiah 11:1 and Micah 6:9). The rod was also an emblem of an apostle's authority in the church (1 Corinthians 4:21).

According to current theories... I am a Levite.

Hebrews 9

Then verily the first covenant had also ordinances of divine service, and a worldly sanctuary.

2 For there was a tabernacle made; the first, wherein was the candlestick, and the table, and the shewbread; which is called the sanctuary.

3 And after the second veil, the tabernacle which is called the Holiest of all;

4 Which had the golden censer, and the ark of the covenant overlaid round about with gold, wherein was the golden pot that had manna, and Aaron's rod that budded, and the tables of the covenant;

YOU HAVE TO HAVE ALL 3 OF THESE!!!
PRAYER BOOKS

REWRITE

Almond in the Strong's Concordance 8246 & 8247. Two Hebrew words that mean almond. Pull up 8244 & 8245, Shaqed is regular almond or being a 'amygdala', while saqad means to be 'on point' or do what the amygdala does. With no nikkud, these are actually the same exact word, same consonants with no vowels. Get it?

Jeremiah says that he sees an almond on a branch or tree, then God says that is alert and vigilant (on point) to perform His Word. This word Almond and 'Alert' is in the Old Testament exactly 10 times. This is again, the self destruct button, if you don't follow the Law, the ten, surely you will die. The Masonic Initiation in Genesis is made plain when you read the Hebrew word as code, spell the word out.

In Hebrew not only can each word be converted into a number, but each letter is own word, making each word a sentence, a broken sentence but a sentence none the less.

Genesis 2

7 And the Lord God formed man of the dust of the

ground, and breathed into his nostrils the breath of life; and man became a living soul.

8 And the Lord God planted a garden eastward in Eden; and there he put the man whom he had formed.

9 And out of the ground made the Lord God to grow every tree that is pleasant to the sight, and good for food; the tree of life also in the midst of the garden, and the tree of knowledge of good and evil.

Notice that man and the trees are formed of the same soil…

16 And the Lord God commanded the man, saying, Of every tree of the garden thou mayest freely eat:

17 But of the tree of the knowledge of good and evil, thou shalt not eat of it: for in the day that thou eatest thereof thou shalt surely die.

Genesis 3

5 For God doth know that in the day ye eat thereof, then your eyes shall be opened, and ye shall be as gods, knowing good and evil.

6 And when the woman saw that the tree was good for food, and that it was pleasant to the eyes, and a tree to be desired to make one wise, she took of the fruit thereof, and did eat, and gave also unto her husband with her; and he did eat.

Again the word "almond" is in the Old Testament ten times, God gave them the Law. The Law was wrapped up in the 'don't eat from that tree' part. This why in this big math book Almond is there ten times, it's the representation of the Ten Commandments or the Law.

שקד

Letter by Letter, the Shin (pronounced sheen looking like a weird w letter) means to eat, consume, or destroy. In the first letter alone you have eat and die. The second letter called Qoff, that looks like a backwards p, means palm of the hand symbolizing to take. The 3rd Letter, Dalet that looks like a backwards r, means doorway. It is used a the doorway between the mortal/spirit realm. The world of the Elohim. The story is like a Russian doll, one inside the other, inside the other, fractal symmetry but a story. The code is Binary too. Good and Evil, evil breaking the law will make the Almond poisonous and give you a Heart Attack. Follow the Law and your heart's desires will be granted, this is Prayer! Literally God will open doors for you, that no man can close! It's basically a short form of repeated law, and there is a bible chapter named 'the Repeated Law' in greek that is deuteronomion or Deuteronomy. There in Deuteronomy is the lay out for what many believe is the story of Black People, being chained and sold around the world. No other people can really make that claim. Every has been in and practiced slavery but no one else was turned into a beast of the field and **a food**, globally exported and imported with receipts... The chapter even says plainly God with make you go crazy shocking your heart!

Deuteronomy 28

1 And it shall come to pass, if thou shalt hearken diligently unto the voice of the Lord thy God, **to observe** and to do all his commandments which I command thee this day, that the Lord thy God will set thee on high above all nations of the earth:

2 And all these blessings shall come on thee, and overtake thee, if thou shalt hearken unto the voice of the Lord thy God.

3 Blessed shalt thou be in the city, and blessed shalt thou be in the field.

4 Blessed shall be the fruit of thy body, and the fruit of thy ground, and the fruit of thy cattle, the increase of thy kine, and the flocks of thy sheep.

5 Blessed shall be thy basket and thy store.

6 Blessed shalt thou be when thou comest in, and blessed shalt thou be when thou goest out.

7 The Lord shall cause thine enemies that rise up against thee to be smitten before thy face: they shall come out against thee one way, and flee

before thee seven ways.

8 The Lord shall command the blessing upon thee in thy storehouses, and in all that thou settest thine hand unto; and he shall bless thee in the land which the Lord thy God giveth thee.

9 The Lord shall establish thee an holy people unto himself, as he hath sworn unto thee, if thou shalt keep the commandments of the Lord thy God, and walk in his ways.

10 And all people of the earth shall see that thou art called by the name of the Lord; and they shall be afraid of thee.

11 And the Lord shall make thee plenteous in goods, in the fruit of thy body, and in the fruit of thy cattle, and in the fruit of thy ground, in the land which the Lord sware unto thy fathers to give thee.

12 The Lord shall open unto thee his good treasure, the heaven to give the rain unto thy land in his season, and to bless all the work of thine hand: and thou shalt lend unto many nations, and thou shalt not borrow.

13 And the Lord shall make thee the head, and not the tail; and thou shalt be above only, and thou shalt not be beneath; if that thou hearken unto the commandments of the Lord thy God, which I command thee this day, to observe and to do them:

14 And thou shalt not go aside from any of the words which I command thee this day, to the right hand, or to the left, to go after other gods to serve them.

15 But it shall come to pass, if thou wilt not hearken unto the voice of the Lord thy God, to observe to do all his commandments and his statutes which I command thee this day; that all these curses shall come upon thee, and overtake thee:

16 Cursed shalt thou be in the city, and cursed shalt thou be in the field.

17 Cursed shall be thy basket and thy store.

18 Cursed shall be the fruit of thy body, and the fruit of thy land, the increase of thy kine, and the flocks of thy sheep.

19 Cursed shalt thou be when thou comest in, and cursed shalt thou be when thou goest out.

20 The Lord shall send upon thee cursing, vexation, and rebuke, in all that thou settest thine hand unto for to do, until thou be destroyed, and until thou perish quickly; because of the wickedness of thy doings, whereby thou hast forsaken me.

21 The Lord shall make the pestilence cleave unto thee, until he have consumed thee from off the land, whither thou goest to possess it.

22 The Lord shall smite thee with a consumption, and with a fever, and with an inflammation, and with an extreme burning, and with the sword, and with blasting, and with mildew; and they shall pursue thee until thou perish.

23 And thy heaven that is over thy head shall be brass, and the earth that is under thee shall be iron.

24 The Lord shall make the rain of thy land powder and dust: from heaven shall it come down upon thee, until thou be destroyed.

25 The Lord shall cause thee to be smitten before thine enemies: thou shalt go out one way against them, and flee seven ways before them: and shalt be removed into all the kingdoms of the earth.

26 And thy carcase shall be meat unto all fowls of the air, and unto the beasts of the earth, and no man shall fray them away.

27 The Lord will smite thee with the botch of Egypt, and with the emerods, and with the scab, and with the itch, whereof thou canst not be healed.

28 <u>**The Lord shall smite thee with madness, and blindness, and astonishment of heart**</u>:

29 And thou shalt grope at noonday, as the blind

gropeth in darkness, and thou shalt not prosper in thy ways: and thou shalt be only oppressed and spoiled evermore, and no man shall save thee.

30 Thou shalt betroth a wife, and another man shall lie with her: thou shalt build an house, and thou shalt not dwell therein: thou shalt plant a vineyard, and shalt not gather the grapes thereof.

31 Thine ox shall be slain before thine eyes, and thou shalt not eat thereof: thine ass shall be violently taken away from before thy face, and shall not be restored to thee: thy sheep shall be given unto thine enemies, and thou shalt have none to rescue them.

32 Thy sons and thy daughters shall be given unto another people, and thine eyes shall look, and fail with longing for them all the day long; and there shall be no might in thine hand.

33 The fruit of thy land, and all thy labours, shall a nation which thou knowest not eat up; and thou shalt be only oppressed and crushed alway:

34 So that thou shalt be mad for the sight of thine eyes which thou shalt see.

35 The Lord shall smite thee in the knees, and in the legs, with a sore botch that cannot be healed, from the sole of thy foot unto the top of thy head.

36 The Lord shall bring thee, and thy king which

thou shalt set over thee, unto a nation which neither thou nor thy fathers have known; and there shalt thou serve other gods, wood and stone.

37 And thou shalt become an astonishment, a proverb, and a byword, among all nations whither the Lord shall lead thee.

38 Thou shalt carry much seed out into the field, and shalt gather but little in; for the locust shall consume it.

39 Thou shalt plant vineyards, and dress them, but shalt neither drink of the wine, nor gather the grapes; for the worms shall eat them.

40 Thou shalt have olive trees throughout all thy coasts, but thou shalt not anoint thyself with the oil; for thine olive shall cast his fruit.

41 Thou shalt beget sons and daughters, but thou shalt not enjoy them; for they shall go into captivity.

42 All thy trees and fruit of thy land shall the locust consume.

43 The stranger that is within thee shall get up above thee very high; and thou shalt come down very low.

44 He shall lend to thee, and thou shalt not lend to him: he shall be the head, and thou shalt be the tail.

45 Moreover all these curses shall come upon thee, and shall pursue thee, and overtake thee, till thou be destroyed; because thou hearkenedst not unto the voice of the Lord thy God, to keep his commandments and his statutes which he commanded thee:

46 And they shall be upon thee for a sign and for a wonder, and upon thy seed for ever.

47 Because thou servedst not the Lord thy God with joyfulness, and with gladness of heart, for the abundance of all things;

48 Therefore shalt thou serve thine enemies which the Lord shall send against thee, in hunger, and in thirst, and in nakedness, and in want of all things: and he shall put a yoke of iron upon thy neck, until he have destroyed thee.

49 The Lord shall bring a nation against thee from far, from the end of the earth, as swift as the eagle flieth; a nation whose tongue thou shalt not understand;

50 A nation of fierce countenance, which shall not regard the person of the old, nor shew favour to the young:

51 And he shall eat the fruit of thy cattle, and the fruit of thy land, until thou be destroyed: which also shall not leave thee either corn, wine, or oil, or the increase of thy kine, or flocks of thy sheep, until he have destroyed thee.

52 And he shall besiege thee in all thy gates, until thy high and fenced walls come down, wherein thou trustedst, throughout all thy land: and he shall besiege thee in all thy gates throughout all thy land, which the Lord thy God hath given thee.

53 And thou shalt eat the fruit of thine own body, the flesh of thy sons and of thy daughters, which the Lord thy God hath given thee, in the siege, and in the straitness, wherewith thine enemies shall distress thee:

54 So that the man that is tender among you, and very delicate, his eye shall be evil toward his brother, and toward the wife of his bosom, and toward the remnant of his children which he shall leave:

55 So that he will not give to any of them of the flesh of his children whom he shall eat: because he hath nothing left him in the siege, and in the straitness, wherewith thine enemies shall distress thee in all thy gates.

56 The tender and delicate woman among you, which would not adventure to set the sole of her foot upon the ground for delicateness and tenderness, her eye shall be evil toward the husband of her bosom, and toward her son, and toward her daughter,

57 And toward her young one that cometh out from between her feet, and toward her children which she shall bear: for she shall eat them for want of all things secretly in the siege and straitness, wherewith thine enemy shall distress thee in thy gates.

58 If thou wilt not observe to do all the words of this law that are written in this book, that thou mayest fear this glorious and fearful name, The Lord Thy God;

59 Then the Lord will make thy plagues wonderful, and the plagues of thy seed, even great plagues, and of long continuance, and sore sicknesses, and of long continuance.

60 Moreover he will bring upon thee all the diseases of Egypt, which thou wast afraid of; and they shall cleave unto thee.

61 Also every sickness, and every plague, which is not written in the book of this law, them will the Lord bring upon thee, until thou be destroyed.

62 And ye shall be left few in number, whereas ye were as the stars of heaven for multitude; because thou wouldest not obey the voice of the Lord thy God.

63 And it shall come to pass, that as the Lord rejoiced over you to do you good, and to multiply you; so the Lord will rejoice over you to destroy you, and to bring you to nought; and ye shall be plucked from off the land whither thou goest to possess it.

64 And the Lord shall scatter thee among all people, from the one end of the earth even unto the other; and there thou shalt serve other gods, which neither thou nor thy fathers have known, even wood and stone.

65 And among these nations shalt thou find no ease, neither shall the sole of thy foot have rest: but the Lord shall give thee there a trembling heart, and failing of eyes, and sorrow of mind:

66 And thy life shall hang in doubt before thee; and thou shalt fear day and night, and shalt have none assurance of thy life:

67 In the morning thou shalt say, Would God it were even! and at even thou shalt say, Would God it were morning! for the fear of thine heart wherewith thou shalt fear, and for the sight of thine eyes which thou shalt see.

68 And the Lord shall bring thee into Egypt again with ships, by the way whereof I spake unto thee, Thou shalt see it no more again: and there ye shall be sold unto your enemies for bondmen and bondwomen, and no man shall buy you.

Crazy right? Obedience comes with blessings. Disobedience comes with Heart Attack! Lets go even crazier, the original Hebrew has no vowels, therefore we can pronounce SH-Q-D as SHOCKED!!! Bruh! Why do people feel the need to mess with the words in the Bible, they speak loud and clear.

SHOCK - 1560s, "violent encounter of armed forces or a pair of warriors," a military term, from French choc "**violent attack**," from Old French choquer "strike against," probably from Frankish, from a Proto-Germanic imitative base (compare Middle Dutch schokken "to push, **jolt**," Old High German scoc "jolt, swing").

The general sense of "**a sudden blow**, a violent collision" is from 1610s. The meaning "a sudden and disturbing impression on the mind" is by 1705; the sense of "**feeling of being (mentally) shocked**" is from 1876.

The electrical sense of "**momentary stimulation of the sensory nerves and muscles caused by a sudden surge in electrical current**" is by 1746. The medical sense of "**condition of profound prostration caused by trauma, emotional disturbance, etc**." is by 1804 (it also once meant "**seizure, stroke, paralytic shock**" 1794).

Shock-absorber is attested from 1906 (short form

shocks attested by 1961); **shock wave** is from 1846. Shock troops (1917), especially selected for assault work, translates German stoßtruppen and preserves the word's original military sense. Shock therapy is from 1917; shock treatment from 1938.

"to come into violent contact; strike against suddenly and violently," 1560s, literal senses now archaic or obsolete, from shock (n.1). The meaning "offend, displease, strike with indignation, horror, or disgust" is by 1650s. The meaning "to give (something) an electric shock" is from 1746. SHOCKED - 1640s, "violently shaken;" 1840, "scandalized," past-participle adjective from shock (v.1).

A HEART ATTACK = HEART SHOCKED, OR DOES THAT JUST MAKE TO MUCH SENSE???

We did the full scientific lay out in the Speak it into Existence book, please read &/or reread that book.

IF YOU FOLLOW THE BLUEPRINT YOU GET BLESSED WITH YOUR WISHES AND MORE IMPORTANTLY **TIME**, THE MAIN THING WE CAN'T BUY! THE SATAN TEAM IS WORKING ON MIMICKING THAT THOUGH, FAKE BODY PARTS AND DRUGS...

THAT WAY MAN CAN GIVE YOU TIME... ONLY SATAN HASNT FIGURED THAT OUT YET, AND CAN NOT, ITS NOT IN HIS NATURE TO NURTURE...

COLD OVER SANITIZED TECHNOLOGY CANT BREATH LIFE INTO THE DIRT THAT YOU ARE, THE SLIME THAT YOU ARE...

ONLY GOD...

HATE IT OR LOVE IT! THE UNDERDOGS ON TOP... IM GONE SHINE HOMEY UNTIL MY _____ STOP... - 50

YOU GOT THE BLUEPRINT ON HOW TO LIVE AND TREAT EACH OTHER....

TREAT - c. 1300, trēten (intrans.), "negotiate, **debate** or **discuss for the purpose of settling a dispute**;" late 14c. as "bargain, deal with;" from Old French traitier "deal with, **act toward**; set forth" in speech or writing (12c.). This is from Latin tractare "**manage**, handle, deal with, **conduct oneself** (in a certain manner) toward," **literally "drag about, tug, haul, pull violently**," frequentative of trahere (past participle tractus) "to pull, draw" (see tract (n.1)). Compare entreat.

The sense of "deal with, handle, or **develop in speech or writing**" is from early 14c.; in reference to tangible objects, "deal with or touch physically," late 14c. The use in medicine "**attempt to heal or cure, to manage in the application of remedies**," is by early 15c. (Chauliac); one of the Middle English senses of treat (n.) was "**medicinal salve**" (late 14c.).
The meaning "**entertain with food and drink without expense to the recipient by way of compliment, good will or kindness**" is by 1710. Related: Treated; treating.

I'm working on a song... this the 'hook' I'm playing

with...
Chorus

Its getting tough naga
I need some love naga
Listen god
I said enough is enough naga
I put my gun down
We at one now
Im eating right working out praising ya sun now

YOU HAVE TO HAVE ALL 3 OF THESE!!! PRAYER BOOKS

DR.ENQI
Raw Herbal Compounds
PRODUCT GUIDE

Detox Kit:

Kemeluminescence -
NRF2, YEAST, FUNGUS, FAT SUPPORT;
Bladderwrack, Yarrow, Cascara Sagrada, Moss, Happy Tree, Madagascar, Periwinkle, Mayapple, Pacific Yew, Cloves, Amla, Coriander, Black Walnut, Kelp, White Pine Bark, Horny Goat Weed, Milk Thistle, Tribulus, Bitter Melon, Chaste Berry, African Pygeum, Cinnamon, Gynesylvestre, Hemp, Pau D Arco, African Bird Pepper, Cinchona Bark, Chinese Senega Root, Biden Pilosa, Houttuynia, Licorice, Skullcap, Scute Root, Ginseng, Rehmania, Er Bu Shir Tao, Bugleweed

Swadj Momatomix -
Marrow & Electromagnetism Support / Rich in Hydrogen, Phosphorus, Aromatic Amino Acid
Phosphorus, Nettles, Wild Lettuce, Hydrogen, Plant Enzyme, &Alkaloid+ MATRIX

Antiviral Kit -

Antiviral
Antifungal
Antibacterial
mtDNA Protector
The most comprehensive organic antiviral kit ever assembled to fight viral infection and improve recovery

Antivirals -
Exogenous & Endogenous Pathogen Support
Cilantro, Celery, Chaparral, Olive Leaf, Oregano Leaf, Black Walnut, Lysine, Tyrosine, Thyme, Cleavers, Hyssop, Bladderwrack, Ginger

Antiviral Nutrient -
Pathogen Suppression Support
Manganese, Rosemary, Hydrangea, Bilberry, Rhizome Rei

Antiviral Oil -
Immunglobulin & Antibody Support
Oregano, Peppermint, Tea Tree, Cinnamon, Hyssop, Thyme, Clove, Ginger

Calcium -
Muscle & Bone Support
Blood Pressure, Insulin Control, Nerve Function, Muscle Contraction
Kelp, Calcium, Sesame, Cloves

Chromium & Vanadium -
Glucose Tolerance Factor & Eyesight Support

Fat Loss , Insulin Metabolism , Hydration , Muscle Integrity , Energy

Chromium, Fenugreek, Vanadium, Bitter Melon, Gymnema Sylvestre

Copper -

Pigment System Support

Cardiovascular Key, Heart Beat Nutrient, White Blood Cell Reg

Copper, Cilantro, Cloves, Milk Thistle

Iron -

Heme & Magnetism Support

Electron Circulation, Digestive System, Thermogenesis, Brain Power

Iron, Yellow Dock, Stinging Nettles, Chaparral

Magnesium -

Energy & Light Metabolism Support

Muscle Function, Energy, Builds ; Proteins/ Enzymes/Hormones , DNA Repair

Blue Vervain, Burdock, Parsley, Magnesium

Muscle Drip -

Children/Adults Multivitamin & Bone Tendon Compound

Blood Oxygen, Breakdown Lactic Acid, Builds Blood Cells Faster, Cleans Lymphatic System

Elderberries, Cherries, Sea Moss, Stinging Nettles, Horsetail, Lily of the Valley, Bladderwrack, Bromide, Melatonin, Phosphorus, Boron, Calcium, Strontium

Muscle Plants -
Children + Adults Multivitamin & Muscle/Joint Compound
Gout, Autography, Enhanced Healing, Arthritis, Remove Stones
Elderberries, Cherries, Bugleweed, Hombre Grande, Blue Vervain, Chaparral, Ginseng, Rhodiola, Boswellia, Eluethero, Melatonin, Phosphorus, Magnesium

Selenium -
Immune Plasma Support
Thyroid Health, Cancer Suppression, Mental Health, Tumor Suppression
Selenium, Burdock, Bladderwrack, Sarsaparilla

Swadj Momatomix -
Marrow & Electromagnetism Support / Rich in Hydrogen, Phosphorus, Aromatic Amino Acid
Phosphorus, Nettles, Wild Lettuce, Hydrogen, Plant Enzyme, &Alkaloid+ MATRIX

Zinc -
Skin & Enzyme Support
Anabolic Boost, Immune System Nutrient, Stem Cell Health, Gene Support
Rosemary, Chlorella, Sage, Zinc

Watermelanin -
Nootropic, Dopamine, Muscle Recovery, Nourish Your Pineal Gland, DMT Support
Raw Organic Non-GMO Black Watermelon Seeds

Lupulin

Anabolic Hormone Help -
Anabolic Hormones, AMPK & Circadian Support
Jiaogulan, Wild Lettuce, Tribulus, Longjack, Maca
& Pollen Blend

Historic -
Histone Sirtuin Support
Grape Skin, Resveratrol, Tyrosine Analogue,
Japanese Knotweed

Ocean Steak -
Vegan B12, Carbon, Nucleoside, Protein,
Nucleotide, Omega 3 & Eye Support
Phytoplankton, Duckweed, Chlorella, Purple
Laver, Chondrus Crispus & C60 Lutein,
Zeaxanthin, Ocean Pigment Matrix

Chrondris Crispus -
Structured Water Mucus Membrane Support
Copper, Cilantro, Cloves, Milk Thistle

NON GMO Moringa -
Whole Body Nutrition Support
Raw Organic Non-GMO Moringa

Purple Phaze -
Anti-Aging Longevity Support
FoTi, Pumpkin Seed, NMN, Bhringaraj, Biotin,
Silica, Tyrosine, Yucca, White Willow Bark, French
Lilac, NAD

Every item on this list, every compound is not only

from God but works on the skin from the inside out, what we need now is topical.

Topical = Tropical

Batana is Great but it's expensive and incomplete.

Researchers identify 135 new melanin genes responsible for pigmentation

Date: August 11, 2023

Source: University of Oklahoma

Summary: The skin, hair and eye color of more than eight billion humans is determined by the light-absorbing pigment known as melanin. New research has identified 135 new genes associated with pigmentation. Vitamin D and Vitamin A... I told you nature doesn't wait to be discovered before getting to work! Melanin vs Diabetes as a Ministry & Movement isn't waiting around to save lives... We have saving lives and creating thought leaders for 20 years!!! The thing is science just discovered 135 genes for pigment and melanin, how the F@3$ have the been acting as if.... This is why we rely on Nature, God & our Ancestors.

We are teaching the world, showing the world...

#HealingLooksLikeThis

Most people don't know what the process of Healing actually Looks Like!!!

People judge health by how your skin looks, literally your complexion. Your Complex of Ions!

Complexion - the general aspect or character of something; the natural color, texture, and appearance of a person's skin, especially of the face.

complexion (n.)
mid-14c., complexioun, "temperament, natural disposition of body or mind," from Old French complexion, complession "**combination of humors**," hence "temperament, character, make-up,"   from Latin complexionem (nominative complexio) "combination" (in Late Latin, "physical constitution"), from complexus "surrounding, encompassing," past participle of complecti "to encircle, embrace," in transferred use, "to hold fast, master, comprehend," from com "with, together" (see com-) + plectere "to weave, braid, twine, entwine," from PIE *plek-to-, suffixed form of root *plek- "to plait."
The Middle English sense is from the old medicine notion of bodily constitution or general nature resulting from blending of the four primary qualities (hot, cold, dry, moist) or humors (blood, phlegm, choler, black choler). The specific meaning "**color or hue of the skin of the face**" developed by mid-15c. In medieval physiology, the color of the face was believed to be

caused by the balance of humors in the body and indicate temperament or health. The word rarely is used in the sense of "state of being complex." also from mid-14c.

Humor - the quality of being amusing or comic, especially as expressed in literature or speech; a mood or state of mind. **Each of the four chief fluids of the body (blood, phlegm, yellow bile (choler), and black bile (melancholy)) that were thought to determine a person's physical and mental qualities** by the relative proportions in which they were present.

humor (n.)
mid-14c., "**fluid or juice of an animal or plant**," from Old North French humour "liquid, dampness; (medical) humor" (Old French humor, umor; Modern French humeur), from Latin umor "body fluid" (also humor, by false association with **humus "earth"**); related to umere "**be wet**, moist," and to uvescere "become wet" (see humid). In old medicine, "any of the four body fluids" (blood, phlegm, choler, and melancholy or black bile).

*The human body had four humors—blood, phlegm, yellow bile, and black bile—which, in turn, were associated with particular organs. Blood came from the heart, phlegm from the brain, yellow bile from the liver, and black bile from the spleen. Galen and Avicenna attributed certain elemental qualities to each humor. Blood was hot and moist, like air; phlegm was cold and moist, like water; yellow bile was hot and dry, like fire; and black bile was cold and dry, like earth. In effect, the human body was a microcosm of the larger world. [Robert S. Gottfried, "The Black Death," 1983]

Their relative proportions were thought to determine physical condition and state of mind. This gave humor an extended sense of "mood, temporary state of mind" (recorded from 1520s); the sense of "amusing quality, funniness, jocular turn of mind" is first recorded 1680s, probably via sense of "whim, caprice" as determined by state of mind (1560s), which also produced the verb sense of "indulge (someone's) fancy or disposition." Modern French has them as doublets: humeur "disposition, mood, whim;" humour "humor." "The pronunciation of the initial h is only of recent date, and is sometimes omitted ..." [OED].

For aid in distinguishing the various devices that tend to be grouped under "humor," this guide, from Henry W. Fowler ["Modern English Usage," 1926] may be of use:

HUMOR: motive/aim: discovery; province: human nature; method/means: observation; audience: the sympathetic
WIT: motive/aim: throwing light; province: words & ideas; method/means: surprise; audience: the

intelligent

SATIRE: motive/aim: amendment; province: morals & manners; method/means: accentuation; audience: the self-satisfied

SARCASM: motive/aim: inflicting pain; province: faults & foibles; method/means: inversion; audience: victim & bystander

INVECTIVE: motive/aim: discredit; province: misconduct; method/means: direct statement; audience: the public

IRONY: motive/aim: exclusiveness; province: statement of facts; method/means: mystification; audience: an inner circle

CYNICISM: motive/aim: self-justification; province: morals; method/means: exposure of nakedness; audience: the respectable

SARDONIC: motive/aim: self-relief; province: adversity; method/means: pessimism; audience: the self

also from mid-14c.

We alway have to remember the Greeks were educated by the Egyptians, ie...

Eu - Good, Perfect, Complete

Melanin (Melanos) - Black

EuMelanin is the new way to reference Osiris.

The new usage of humor came from state of mind, which came from your health, based in the balance of fluids.

#StayWet

The herbs and bitters are to condition the internal fluids, the BleuMagick conditions the body's waters, we are now going to top it off with #SkinFood, to make sure you can #StayWet.

Please read &/or reread Lymphatic Immunity, Mitochondria Water, HydroChemistry...

You've got to learn everything you can from these books about Water. Then you will be ready to apply these principles and practices, Kitchen Chemistry, Orthorexia & this book espouse! Truth be told, these 3 books are plug and play immediately... but the further study is what refines you. You have to disappear sometime and come back stronger. You don't that by consuming new information in your absence.

The biggest difference between those in the Rat Race, and those who aren't, is Priority of Consuming New Information. Reading.

What does it mean to you that, the other pigments in the skin control Melanin production?

What does it mean to you that, they just discovered 3000 new types of Neurons?

Neurons are either specialized Melanocytes or Melanocytes are specialized Neurons, they have discovered 3,000 new types Naga.... Wake Up!!!

Either your whole body is a Brain or a Heart! The Heart has all of these cells, Neurons, Nerves, Melanocytes etc...

What does it mean to you that, they just discovered 135 genes that are associated with Pigment? Is that 135 genes for the Skin? Brain? Heart???

My goal is to make #SkinFood inexpensive enough for you and your family to use twice a day, that way we can #StayWet.

In L'Goat we explain in detail how water builds what it needs, with the right conditions. First thing is well, Water. Second thing obviously is Retinoids, to fertilize the soil. Then of course Sunlight...

You are the Dust of the Ground, Divine Soil, remember that the ground exerts pressure on seeds. You need the proper exercise, to create that mechanical pressure to stimulate growth and regeneration. Mechanical Pressure helps to circulate Magnetism, this is the #BleuMagick of #ElectroMagneticTissue.

If this is your first book of ours that you have read... God Bless Eu, or maybe this isn't your first book but you haven't read PiezoElectroChemistry, please do that...

You are the Fruit of Melanin, your S.elf O.rganizing U.niversal L.ight.

Soul is the Fruit of PhotoVoltaic Pigment.

YOU HAVE TO HAVE ALL 3 OF THESE!!!
PRAYER BOOKS

TECHNOLOGY EQUIPMENT RECAP

HydroGenes - 1! Proton and 1! Electron.

Melanocytes - Photovoltaic Cells

Neurons/Nerves - Electromagnetic Cells

Fascia - Plasma Medium (misonomered Ether)

Brain - CPU, Inductor

Brainstem - Two-way Adapter for CPU into the Motherboard

Pineal Gland - Receiver, Crystal Tuner & Actuator Arm/Head responsible for Phosphorescence, Thermoluminescence, Piezoelectricity, Birefringence & Harmonic Generation (very much like the otoconia in the ears)

Operating System - Deductive Logic or PQ

Heart - Hydraulic Ram, Turbine (from the Greek τύρβη, tyrbē, or Latin turbo, meaning vortex) and Hard Drive.

Melanosomes - Alternators

Mitochondria - Motors

Myelin Sheath - Insulation

Cytoskeleton - Filaments

Phospholipids - Capacitors, Dielectric Material (lipids in general)

Spine - Piezoelectric, Motherboard, Radio Wave Antenna

RBC - Floppy Discs

Lymph Nodes - Filters, Nodes

Tastebuds - Electronic Scanners

Protein - Transformer

Transformers 'Roll Out' - Conformational Change (Macromolecule Shape Shifting)

Antioxidants - Semi-conductors (especially the selenium based...)

Body Cells - Plasma based Crystal disc, fitted with integrated circuits as well as gates and channels (see Human Cell Membrane and/or Computer Chip)

Nerves & Vessels - 'Copper' wires (CoAxial Cables) and Fiber Optics

Pigment, Nerve & Blood Clusters - Input Devices like a Mouse, Keyboard, etc…

DNA - Piezoelectric, Antenna, Data Storing Inductors.

DNA sub entry **Tissues** - Short Living Stories.

DNA sub entry **Genome** - Substrate & Product, a digital Library (HardDrive) of all your Ancestors have ever seen, said, touched, tasted or heard.

DNA sub entry **Chromosome** - Rewritable Unlimited Storage Books (Folders) of the Library, DNA.

DNA sub entry **Histone** - Writing instrument, encoders and **book spines**.

DNA sub entry **non-coding RNA** - Self Organizing Books Shelves

DNA sub entry **Gene** - Chapters (Files) in the Books, source codes.

DNA sub entry **Messenger RNA** - Protein Information, a Sentence.

DNA sub entry **Codon** - Word (Binary Code there are 2 bonds between each 3 nucleotides representing their arrangement), Amino Acid.

DNA sub entry **Nucleotide** - Letter

DNA sub entry **Nucleoside** - Bits of Information

Collagen Based Tissue - Piezoelectric Inductors

Stomach - Chemical Mixer

Lumen - the SI unit of luminous flux = to the amount of light emitted per second….. or hollow structures in vessels and cells… hmmm????

Eyes - Camera Lens/Charge Coupled Device (CCD), Digital to Analogue Converter, Complex Photovoltaic Cells/Photodetector…

Amino Acids - Fuses that can be almost anything!

Nucleic Acid - Actual Intelligence (self powering too).

N-Type Semiconductors - Selenium or Silica doped with Phosphorus (Alkaline-ish)

P-Type Semiconductors - Selenium or Silica doped with Boron (Acid-ish)

PN Junction - Crystal Lattice Structure Material allowing the flowing of electrons in one direction.

Bone and Fascia seem to be a massive N-Type, P-Type, PN Junction Super computer on it's own… especially if we add in the Piezoelectricity & Vitamin D!

Rectifier - N-Type + P-Type + PN Junction in Bone

Melanin - CPU Core, Solar Repeater

Human Cell Membrane and/or Computer Chip - A flat semiconducting (crystal) disc or wafer, with integrated circuits (resistors/conductors) and/or gates & channels. We now have to add the filaments into this Crystal Disc we call a Body Cell or Somatic Cell.

Transistors - a semiconductor device with three connections, capable of amplification in addition to rectification.

The location that a virus goes viral in, is called a **Hotspot**? WTH!

Virus - an infective agent that typically consists of a nucleic acid molecule in a protein coat, is too small to be seen by light microscopy, and is able to multiply only within the living cells of a host.

Wait you see that, it is happening again! Host...

See look there is another definition of **Virus** - a piece of code that is capable of copying itself and typically has a detrimental effect, such as corrupting the system or destroying data.

Wait a damn minute! DNA is a piece of **code**... A viral strand of DNA or RNA that can jump host is fully capable in that context of copying itself, one would even argue, that is it's only 'motion'.

The detrimental effects of corrupting the system (physical illness) or destroying data (mental illness), can clearly be seen anthropomorphically.

Alien - Virus

Culture - the arts and other manifestations of human intellectual achievement regarded collectively, the customs, arts, social institutions, and achievements of a particular nation, people, or other social group. The cultivation of bacteria, tissue cells, etc. in an artificial medium containing nutrients, a preparation of cells obtained from a culture. The cultivation of **plants**.

Going Live - become operational.

Live Stream - a live transmission of an event over the internet, transmit or receive live video and audio coverage of (an event) over the internet.

Download - copy (data) from one computer system to another, typically over the internet, an act or process of downloading data.

Upload - transfer (data) from one computer to another, typically to one that is larger or remote from the user or functioning as a server, an act or process of downloading data.

Data - facts and statistics collected together for reference or analysis; the quantities, characters, or symbols on which operations are performed by a computer, being stored and transmitted in the form of electrical signals and recorded on magnetic, optical, or mechanical recording media.

Host - an animal or plant on or in which a parasite or commensal organism lives. VS

Host - store (a website or other data) on a server or other computer so that it can be accessed over the internet.

Transmission is the act of transferring something from one spot to another, like a radio or TV broadcast, or a disease going from one person to another.

I am highlighting the unknown and proposing we may have some answers! <u>Infection - an infectious disease.</u>

plural noun: infections "a chest infection"

Vs

Infection - the presence of a virus in, or its introduction into, a computer system. What is a computer system?

Computer System - a computer system is a programmable electronic device that can accept input; store data; and retrieve, process and output information.

Pandemic language = Virology/Biology

language. The question is, why? The next question is what does that have to do with Dr. Sebi or Robert Becker? The obvious....

Computer System - a computer system is a programmable electronic device that can accept input; store data; and retrieve, process and output information.

Computer System - a single information processor but usually a group of processors that have specified and general computations; grouped by hardware ie... liver cells, lung cells, brain cells etc.. What you think?

Exercise - activity requiring physical effort, carried out to sustain or improve health and fitness.
"exercise improves your heart and lung power"

Exercise - computer training or computer based training.

Exigenetics - Term created by Dr. EnQi for Melanin vs Diabetes research, denoting the control that exercise has over gene expression.

Hydration - the process of inducing gelling, ionizing, dissolution & turbulent flow with activation of cytochrome c (via infrared light).

Resonance - the quality in a sound of being deep, full, and reverberating. "the resonance of his voice"

- The ability to evoke or suggest images, memories, and emotions."the concepts lose their emotional resonance"

- The reinforcement or prolongation of sound by reflection from a surface or by the synchronous vibration of a neighboring object.

- The condition in which an electric circuit or device produces the largest possible response to an applied oscillating signal, especially when its inductive and its capacitative reactances are balanced.

- The condition in which an object or system is subjected to an oscillating force having a frequency close to its own natural frequency.

- The occurrence of a simple ratio between the periods of revolution of two bodies about a single primary.

- The state attributed to certain molecules of having a structure that cannot adequately be represented by a single structural formula but

is
a composite of two or more structures of higher energy.

•   • A short-lived subatomic particle that is an excited state of a more stable particle.

Induction - the action or process of inducting someone to a position or organization."the league's induction into the Baseball Hall of Fame"

Induction - a formal introduction to a new job or position.plural noun: inductions
"an induction course"
enlistment into military service.

Induction - The process or action of bringing about or giving rise to something."isolation, starvation, and other forms of stress induction" the process of bringing on childbirth or abortion by artificial means, typically by the use of drugs.

Induction - The inference of a general law from particular instances.

Induction -"the admission that laws of nature cannot be established by induction" the production of facts to prove a general statement.

Induction - a means of proving a theorem by showing that if it is true of any particular case it is true of the next case in a series, and then showing that it is indeed true in one particular case.

Induction - noun: mathematical induction; plural noun: mathematicals inductionthe production of an electric or magnetic state by the proximity (without contact) of an electrified or magnetized body.

Induction - The production of an electric current in a conductor by varying the magnetic field applied to the conductor.

Induction - The stage of the working cycle of an internal combustion engine in which the fuel mixture is drawn into the cylinders.

Is there anyone reading this that would disagree with our body fitting these definitions, the definitions of a computer?

Man this thought experiment just got a lot more interesting didn't it? MIT and the US Military are different types of receipts huh? Is it possible frequency resonance, spreads disease? Human modems? Can Shedding be a broadcast signal?

Wi-Fi is a wireless networking technology that uses radio waves to provide wireless high-speed Internet access. A common misconception is that the term **Wi-Fi** is short for "wireless fidelity," however Wi-Fi is a trademarked phrase that refers to IEEE 802.11x standards.

Viral shedding is a term for when viruses are replicating or reproducing, the virus is being led out of the host cell where it's replicating or reproducing ... Viral shedding is the expulsion and release of virus progeny following successful reproduction during a host cell infection. Once replication has been completed and the host cell is exhausted of all resources in making viral progeny, the viruses may begin to leave the cell by several methods.

Vaccine - a substance used to stimulate immunity to a particular infectious disease or pathogen, typically prepared from an inactivated or weakened form of the causative agent or from its constituents or products.

Vaccine - a program designed to detect computer viruses and inactivate them.

"the rate of use of vaccines for computer viruses is not as high as in the US, Japan, and other countries"

Application - a medicinal substance put on the skin.

Application - a program or piece of software designed and written to fulfill a particular purpose of the user.

In our thought experiment, if a virus is simply the media for harmful information...

Media - an intermediate layer in the wall of a blood vessel or lymphatic vessel.

Media - the main means of mass communication (broadcasting, publishing, and the internet) regarded collectively.

DOPE - an illicit drug (such as heroin or cocaine) used for its intoxicating or euphoric effects especially : MARIJUANA (dopamine altering)

Dope - a preparation (such as an anabolic steroid, diuretic, or tranquilizer) given to a racehorse to help or hinder its performance

To Dope - In semiconductor production, to dope is the intentional introduction of impurities into an intrinsic semiconductor for the purpose of modulating its electrical, optical and structural properties. The doped material is referred to as an extrinsic semiconductor.

Short Circuit - Cardiac Arrest?

Short Circuit - Multiple Sclerosis (due to loss of insulation)

Overheating - Fever?

Overcurrent - Inflammation

With Infection and Virus included we are onto something.

Current - belonging to the present time;

happening or being used or done now.

Current - <u>a body of water</u> or air <u>moving in a *definite* direction</u>, especially <u>through a surrounding body of water</u> or air in which there is less movement.

Current - a flow of electricity that results from the ordered directional movement of electrically charged particles.

Current - a quantity representing the rate of flow of electric charge, usually measured in amperes.

Current - the general tendency or course of events or opinion.

Leakage Current - the unintended loss of energy, gain of resistance or results of faulty/worn out insulation.

Plasma - Electric Currents or Electric Current Carrier

Electric Current - Magnetic Field (AtomSphere) Carrier

Alternating Magnetic & Electric Waves - Light

NeuroTransmitters - Record of ElectroMagnetic Waves produced by Neurons (ElectroChemical Message)

Hormones - Large simple versions of

NeuroTransmitters (ElectroChemical Message)

Malware - External Negative Mental Programming

Food - Informative Electronic Batteries

0) Movement and sound create energy from water for basic cellular function, via the EnQi Cycle which includes Mitochondria Water. This system slowly increases as all other energy systems fail.

1) Phosphocreatine - anaerobic (no respiration required), phosphocreatine donates it "phospho" to ADP to recycle ATP. This makes 10 ATP per second, its a 1 to 1 ratio (1 phosphocreatine creates 1 ATP) and this is the jump start energy.

2) Anaerobic Glycolysis - anaerobic (no respiration required), Glycogen &/or Glucose to Lactate, 5 ATP per second, 1 to 3 ratio (1 Glycogen creates 3 ATP while 1 Glucose creates 2 ATP) and this is bulk of the energy we focus on, 9 - 120 seconds.

3) NAD/Cytochrome 1 - aerobic (requires oxygen), Glycogen &/or Glucose to CO_2/H_2O, 2.5 ATP per second, 1 to 38 ratio (1 Glycogen &/or Glucose creates 38 ATP), 2 minutes up to 2 hours.

4) FAD/Cytochrome 2 - aerobic (requires oxygen), FFA &/or Triglycerides to CO_2/H_2O, 1.5 ATP per second, 1 to 360 ratio (1 Glycogen &/or Glucose creates 360 ATP), 2 minutes up to 2 days.

Food rule of thumb - Resynthesis of ATP of Inverse

to Yield, the closer the ratio is to 1:1 the fast it can be recycled.

Muscle rule of thumb - Frequently used muscle is slow twitch, Fast twitch is slowly used (at that's the blueprint).

Electric Power - the **rate** at which work is done or energy is transformed into an electrical circuit. Simply put, it is a measure of how much energy is used in a span of time.

Conductor - a person who directs the performance of an orchestra or choir.

Conductor - a material or device that conducts or transmits heat, electricity, or sound, especially when regarded in terms of its capacity to do this.

Lymphatic System - Watermill

Circulatory System - Generator

Integumentary System - Photovoltaic Diaphragm

Immune System - Antivirus, Malware Scanner, Frequency Filter & Rectifier

Nervous System - Power Transmission and Cellular Communications Lines

Fascia System - HydroElectric Grid, Scaffolding

Respiratory System - Windmill

Windmill - a structure that converts wind power or "air" power into rotational energy or vortex energy, to mill grain. In our case grain is Magnetism!

MAGNETS ARE DEFINED BY GRAINS
MAGNETIC GRAINS ARE DEFINED BY APPLIED
STRESS AND CRYSTAL GEOMETRY
SPM SUPERMAGNETIC
SD SINGLE DOMAIN
PSD PSEUDO DOMAIN
MD MULTIDOMAIN

Reproductive System - Quine (self-replicating programs)

Skeletal System - Piezoelectric Crystal Shaped to produced highly specific frequency under stress, Dynamic Oscillators.

Urinary System - Industrial Wastewater, Return Flow, Surface Runoff, Urban Runoff Agricultural & Animal Husbandry Wastewater

Digestive System - Massive Inductor

Mouth - Industrial Grinder

Endocrine System - Programmer for Human Cell Membrane and/or Crystal Gel Computer Chips

Human Being - Resonator

Vessels - Pipes

ELECTRICIAN PRAYER MANUAL

Aromatic Ring - Cyclotron (Particle Accelerator)

Glycation - Corrosion

Exegenetics - Holistic Biomechanics; the purposeful science of combining light, water, diet & exercise to effect DNA.

EnQi's 1st Law of Metabolism - The conversion rate of cholesterol should match the activity of Melanin in the skin. These two systems are designed to be and stay coupled. A dark skin person with low sunlight intake and low exercise is going to die from a Metabolic Complication. The only time Animal Flesh is safe to be consumed by a Eumelanin Dominant person is in times of starvation or extremely high activity.

This Law is a Constant and when broken results in disease every time.

EnQi's 2nD Law of Metabolism - The average rate of applied mechanical stress on the bone electrically stimulating bone marrow, determines the rate of bone deterioration and Red Blood Cell production.

EnQi's 3rd Law of Metabolism - The human body metabolizes Transverse Waves and Mechanical Waves into Electricity. Electricity is the main driver of Biochemistry. Exercise is just as potent a driver of Biochemistry as the Sun.

EnQi's 4th Law of Metabolism -

Electron movement and bonding is the Nature of Chemistry. PhotoChemistry and PiezoElectroChemistry are the Primary drivers of Biochemistry.

The Ancients discovered this and created Martial Artforms as a way to clean the Bone, Bone Marrow & Brain. Plaque & Sugar are the top drivers of Brain Disease. The things destroying the Heart are secondarily destroying the brain, and they are the breaking of these Universal Laws.

EnQi's 5th Law of Metabolism - Nutrients are actually substrates that must be transformed via biochemistry to be meaningful. This means that providing your body with lots of nutrition without the Water, Light & Exercise don't work alone.

EnQi's 6th Law of Metabolism - The Body maintains the least amount of bone marrow required to handle blood demand. The marrow is very energy demanding, thus attracting and storing fat for energy, eventually becoming fat itself. Fatty bone marrow is called yellow bone marrow. Yellow Bone Marrow can be reconverted to Red Bone Marrow should the body's demands require it, and the body's resources facilitate it. The primary driver is pressure, hormesis training on the Bones. BMR is heavily driven by Bone Marrow, this means Bone Marrow is a driver if insulin and insulin resistance.

EnQi's 7th Law of Metabolism - The system of pigments throughout the body are for metabolism of Light, actual Soulfood. The Adsorption & Absorption of Photons by Water.

Adsorption - increase in the concentration of a dissolved substance at the interface of a condensed and a liquid phase due to the operation of surface forces.

Absorption - a physical or chemical phenomenon or a process in which atoms, molecules or ions enter some bulk phase – liquid or solid material. This is a different process from adsorption, since molecules undergoing absorption are taken up by the volume, not by the surface.

EnQi's 8th Law of Metabolism - in a diabetic state, sugar is simply invisible to the body. Sugar is not being "sensed" because it's not being converted to energy. In this state of starvation the body turns on every pathway it has to produce sugar from everything you have in your body, fats and proteins included.

This is the reason that it seems like no matter what you eat or 'don't eat', your blood sugar goes up. It's very frustrating. The only way to make it stop is converting that substrate (glucose) into it's final product (energy). The reception of the actual energy, tells the body to stop producing substrate, we good. This must start in the legs and back, the largest muscles but most overlooked. The legs are particularly punished by sitting for extended periods of time, 3-6 hours straight, for a total over 3/4 the time your awake! The leg circulation atrophies and destroys the nerves, nerves are neurons that need a lot of nutrients!

*You must cross reference any and all protocols; food, exercise, medication etc... with the Constitution book & Declaration of Independence!

YOU HAVE TO HAVE ALL 3 OF THESE!!!
PRAYER BOOKS

MATH BOOK

This chapter will be short and sweet, almost every bible scholar forgets that the Bible is a Math Book, in Math Geometry is Proof. Geometry is number made flesh, energy made matter. We are going to focus on the light of the world (pun intended).

6th Day, 3 Days, 72 hours - 3,6,9 just a co-inky dink of course.

In scripture, 72 symbolizes completeness or fullness in the context of God's plan for humanity. The 72 disciples mission to all nations. The 72 scholars who translated the Hebrew Bible into Greek. In Kabala, it is believed that there are 72 angels that govern various aspects of the world and are also seen as guardians of nations. Jacob's Ladder in the Old Testament of the Bible (Genesis 28, 11-19) and the Zohar, had 72 steps to the Earth and Heaven with angels traveling up and down the steps.

The Table of Nations: In Genesis 10, there are 72 nations listed that emerged from the descendants of Noah, indicating the expansion and diversity of humankind following the Flood.

The 72 Elders of Israel: In Numbers 11:24-25, Moses gathered 70 elders to assist him, but it is often interpreted in Jewish tradition that with Moses himself included, this totals to 72 or two men, one named Eldad and the other Medad, were not in the gathering but had been left in the camp. They too had been on the list, but had not gone out to the tent; yet the spirit came to rest on them also, and they prophesied in the camp. These elders were endowed with the Spirit to help bear the burdens of leadership.

The number 72 holds profound significance in Kabbalistic tradition, primarily associated with the concept of the 72 Names of God. This mystical interpretation draws from specific biblical verses and the deeper meanings attributed to these names.

The 72 Names of God are derived from three consecutive verses in the Book of Exodus (Exodus 14:19-21), each containing exactly 72 letters. The verses describe the miraculous parting of the Red Sea by Moses. The process to extract these names involves writing the three verses one above the other, with the middle verse written backwards.

Verses Used: The three verses are:

Exodus 14:19: "And the angel of God, who went

before the camp of Israel, removed and went behind them..."

Exodus 14:20: "And it came between the camp of Egypt and the camp of Israel..."

Exodus 14:21: "And Moses stretched out his hand over the sea..."

By taking the first letter of the first verse, the last letter of the second, and the first letter of the third, and following a specific pattern, one can form the 72 unique triads.

Each of the 72 names is a three-letter sequence that is believed to resonate with specific spiritual frequencies. The names encapsulate distinct attributes of God and serve various purposes, such as:

Healing and Protection: Certain names are traditionally invoked for healing, protection, and spiritual assistance.

Transformation and Connection: Kabbalistic teachings suggest that these names can be used in meditations and prayer to transform one's life and to connect with divine energies.

Influencing Reality: Practitioners believe that these names have the power to influence reality and manifest desired outcomes in one's life, essentially serving as tools for metaphysical empowerment.

In Kabbalah, the 72 Names correspond to divine attributes such as Chesed (kindness), Gevurah (severity), and Tiferet (beauty), which together represent the balance of divine forces. The number 72 represents the precession of the equinoxes, that is it represents the numerical sequence linked to the earth's axial precession, which causes the apparent alteration in the position of the constellations one degree every 72 years. It has been noted also that Angkor Wat is located 72 degrees of longitude east of the Giza Pyramids.

I am just highlighting the possibility that there is information associated with these numbers. Particularly the fact that the vortex math or ouroboros in number comes from the Bible. Solomon is also said to have used a mysterious worm tool (Shamir) that could cut stones and gems, I would definitely agree that is a vortex reference but it's not in the Bible so we won't discuss it here.

IN PREVIOUS BOOKS WE DISCUSS THAT 22 LETTERS OVER 7 DAYS GIVES THE NUMBER PI OR 3.14. THERE IS A THING HERE, THE GENESIS VERSE IS 27 LETTERS LONG AND THE NUMBER 27 IS SELF LOCATING IN THE NUMBER! THIS MEANS 27 IS 27 PLACES OVER IN THE NUMBER, ITS CALLED A SELF LOCATING NUMBER!!! THE HEBREW ALPHABET IS A SELF CREATING NUMBER **CIPHER** AND THE BIBLE IS A **MATH**

BOOK THAT ENCODES THE MATH OF KHEPRI, THE SELF CREATED GOD THAT MOSES LEARNED ABOUT BEING TAUGHT BY THE EGYPTIANS.

3.141592653589793238462643383<u>27</u>

Cipher - late 14c., "arithmetical symbol for zero," from Old French *cifre* "nought, zero," Medieval Latin *cifra*, which, with Spanish and Italian *cifra*, ultimately is from Arabic *sifr* "zero," literally "empty, nothing," from *safara* "to be empty;" a loan-translation of Sanskrit *sunya-s* "empty." Klein says Modern French *chiffre* is from Italian *cifra*.

The word came to Europe with Arabic numerals. From "zero," it came to mean "any numeral" (early 15c.), then (first in French and Italian) "**secret way of writing; coded message**" (a sense first attested in English 1520s), because early codes often substituted numbers for letters. Meaning "**the key to a cipher or secret writing**" is by 1885, short for cipher key (by 1835).

Figurative sense of "something or someone of no value, consequence, or power" is from 1570s.

also cypher, 1520s, "to do arithmetic" (with Arabic numerals), from cipher (n.). Transitive sense "reckon in figures, cast up" is from 1860. Meaning "**to write in code or occult characters**" is from 1560s. Related: Ciphered; ciphering.

Literally the fact that every Hebrew Letter is a word and a number, means anything written in

Hebrew is a Cipher or Code!

Math is the Code for Energy, Music and Geometry. Math, Energy, Music and Geometry encapsulates all knowledge. All that said to say what? Your body is God's Temple, and that Temple is a Complex Electronic System designed to Transmit and Receive Data, from your creator to carry out your specific tasks in Life.

Here is a new piece for the family. The Kissing Code:

Kiss - Old English coss "a kiss, **embrace**," noun derived from kiss (v.). It became Middle English cos, cus, but in Modern English this was conformed to the verb.

Meaning "small chocolate or candy piece" is from 1825; compare Shakespeare's kissing comfits (1590s) in reference to little sweets used to freshen breath. Kiss-proof, of lipstick, is from 1937. Kiss of death in figurative sense "thing that signifies impending failure" is from 1944 (Billboard magazine, Oct. 21), ultimately in reference to Judas's kiss in Gethsemane (Matthew xxvi.48-50). The kiss of peace was, in Old English, sibbecoss (for first element, see sibling).

Old English cyssan "to touch with the lips" (in respect, reverence, etc.), from Proto-Germanic *kussjan (source also of Old Saxon kussian, Old Norse kyssa, Old Frisian kessa,

Middle Dutch cussen, Dutch, Old High German kussen, German küssen, Norwegian and Danish kysse, Swedish kyssa), from *kuss-, probably ultimately imitative of the sound. Gothic used kukjan. Of two persons, "to reciprocally kiss, to kiss each other," c. 1300. Related: Kissed; kissing. The vowel was uncertain through Middle English; for vowel evolution, see bury.

Kissing, as an expression of affection or **love**, is unknown among many races, and in the history of mankind seems to be a late substitute for the more primitive rubbing of noses, sniffing, and licking. The partial agreement among some words for 'kiss' in some of the IE languages rests only on some common expressive syllables, and is no conclusive evidence that kissing was known in IE times. [Carl Darling Buck, "A Dictionary of Selected Synonyms in the Principal Indo-European Languages," 1949]

A common ku- sound may be found in the Germanic root and Greek kynein "to kiss," Hittite kuwash-anzi "they kiss," Sanskrit cumbati "he kisses." Some languages make a distinction between the kiss of affection and that of erotic love (compare Latin saviari "erotic kiss," vs. osculum, literally "little mouth"). French embrasser "kiss," but literally "embrace," came about in 17c. when the older word baiser (from Latin basiare) acquired an obscene connotation.

To kiss the cup "drink liquor" is early 15c. To kiss the dust "die" is from 1835. To kiss and tell is from 1690s. Figurative (and often ironic) kiss (something) goodbye is from 1935. To kiss (someone) off "dismiss, get rid of" is from 1935, originally of the opposite sex. Insulting invitation kiss my arse (or ass) as an expression of contemptuous rejection is from at least 1705, but probably much older (see "The Miller's Tale").

Respectfully, the knowledge in the Bible is endless, and should not push people away but bring them closer.

Here are the relevant stages throughout the text to this particular code.

1)   Adam & Eve - Ignorant
2)   Solomon - Enlightenment
3)   Jesus - Sharing enlightenment with the world

<u>It is very important to know that the cause and cure for heart disease is located in the Bible</u>. Our goal is to save lives and bring people to patient, diligent, open heart study of the Bible.

1 Corinthians 3

18 Let no one deceive himself. If anyone among you seems to be wise in this age, let him become a fool that he may become wise. **19** For the wisdom of this world is foolishness with God. For it is written, "He catches the wise in their *own* craftiness"; **20** and again, "The Lord knows the thoughts of the wise, that they are futile."

This is speaking specifically to those, 'I have this degree and that degree folks' that use their degrees to change the meaning of the words in the Bible. A degree does not qualify you to rewrite the Bible or tell people that you can't take the words of the Bible seriously or literally. They want to transform the words because they have been trained to do so. You see the old school game was that no one could read the written word of God. The only the wealthy and slave masters, well now that anyone can pick up a bible THEY ARE CHANGING EDITS LIKE A WOMAN WHO HIT THE LOTTO CHANGE SHOES!!! Then to double down, they training folks to tell you that you need to be in the spirit to read the Bible. Then to add insult to injury, if you ask them how are they qualified they tell you about their degrees from college. If P then Q, that means colleges are now Gods that somehow are activating Holy Spirits in these streets??!! GTFOH!!!

Start asking these guys for transcripts, then look up every teacher for every credit, THESE BUFFOONS & COONS getting credits from atheists, agnostics, gay people, Muslims, drug addicts, PDfiles and all manors of sinners... How then does a motley crew of sinners active your Spirit? Easy, for Satan. They gotta stop playing out here! Now where were we?

Here are the relevant stages throughout the text to this particular code.

1)   Adam & Eve - Ignorant
2)   Solomon - Enlightenment
3)   Jesus - Sharing enlightenment with the world

The word is made flesh literally through water, notice in the Bible water unlocks the Holy Spirit (its at least the conduit). I think after years of research that everyone should be baptized, now by who, that's a personal choice. I do not think it should be random though. People are googling baptism near me, like nah bruh! Anyway...

I got the nickname Founder of the Soul (like FOT instead of Goat), by showing scientifically we live on light, not sugar and not ATP. For me it was low hanging fruit, all textbooks tell you that ATP is released into the cytoplasm (water) and then oxidized into ADP. The oxidizing reaction is exothermic, that energy released powers the cell. That right there should've been enough but you know our people, so I just adding for reference the natural behavior of Phosphorus, it releases light in the presence of oxygen. Done!

To create humans we start with 2 cells, please go read &/or reread our book Let there be Music! We document the divine spark of life and the moment (plus the science) of conception. It starts with 2 cells though, kissing. The bible even references Adam's rib for you, another dagger! It wasn't know until recently that in Adults only their ribs and spine have active Stem Cells.

Two is the first number, 2 and a hidden 1 to be exact. Which is a hot debate, who was the Naggar talking to Eve? A actual snake, a dragon, the devil etc...

The second stage of development for the chosen bloodline is Self Development. You see the trick to Self Development is there is no Self! You are never alone, nothing you do in life only affects you, hence the phrase do unto others...

Self Development or the journey to enlightenment ends with Solomon. Solomon reaches the apex of Knowledge and Wisdom. Key note, Solomon became Egyptian before receiving the Wisdom via marriage (then asked to be able to properly administer Maat), yet another reason we need to stop the division.

1 Kings 3

Solomon made an alliance with Pharaoh king of Egypt and married his daughter. He brought her to the City of David until he finished building his palace and the temple of the Lord, and the wall around Jerusalem. **2** The people, however, were still sacrificing at the high places, because a temple had not yet been built for the

Name of the Lord. **3** Solomon showed his love for the Lord by walking according to the instructions given him by his father David, except that he offered sacrifices and burned incense on the high places.

4 The king went to Gibeon to offer sacrifices, for that was the most important high place, and Solomon offered a thousand burnt offerings on that altar. **5** At Gibeon the Lord appeared to Solomon during the night in a dream, and God said, "Ask for whatever you want me to give you."

6 Solomon answered, "You have shown great kindness to your servant, my father David, because he was faithful to you and righteous and upright in heart. You have continued this great kindness to him and have given him a son to sit on his throne this very day.

7 "Now, Lord my God, you have made your servant king in place of my father David. But I am only a little child and do not know how to carry out my duties. **8** Your servant is here among the people you have chosen, a great people, too numerous to count or number. **9** So give your servant a discerning heart to govern your people and to distinguish between right and wrong. For who is able to govern this great people of yours?"

10 The Lord was pleased that Solomon had asked for this. **11** So God said to him, "Since you have asked for this and not for long life or wealth for yourself, nor have asked for the death of your enemies but for discernment in **administering justice**, **12** I will do what you have asked. **I will give you a wise and discerning heart**, so that there will never have been anyone like you, nor will there ever be. **13** Moreover, I will give you what you have not asked for—both wealth and honor—so that in your lifetime you will have no equal among kings. **14** And if you walk in obedience to me and keep my decrees and commands as David your father did, I will give you a long life." **15** Then Solomon awoke—and he realized it had been a dream.

The lord make Solomon wise through his Heart not Brain! For me remember proof is Math for Energy, Music, Geometry etc…

2 Samuel 22 (also Psalms 3:3 interesting the 6 motif)

1 Then David spoke to the Lord the words of this song, on the day when the Lord had delivered him from the hand of all his enemies, and from the hand of Saul.
2 And he said: "The Lord *is* my rock and my fortress and my deliverer;
3 The God of my strength, in whom I will trust;
My shield and the horn[a] of my salvation,
My stronghold and my refuge;
My Savior, You save me from violence.

4 I will call upon the Lord, *who is worthy* to be praised;
So shall I be saved from my enemies.
36 "You have also given **me the shield of Your salvation**; Your gentleness has made me great.

Legend has it that the Star is a symbol of the Crocus Flower or vice versa, for me the 6 points was important. You start out the Bible with 6 days of creating life, and guess what? Life could not exist without the 6 or Hexagram. The Hexagram represents the knowledge of all living things. The Structured Water inside cells, the pigment arrangement in your eyes, the list is too long. In short sphere packing, the science of surface chemistry and tissue forming combine into the hexagonal micro-structures we call flesh.

Here are the relevant stages throughout the text to this particular code.

1) Adam & Eve - Ignorant
2) Solomon - Enlightenment
3) Jesus - Sharing enlightenment with the world

OK now, we get to Jesus, wait... The first number code is 2 & 1, then the second one 6 and 1, the 3rd is 12 and 1. There is no more well known motif for this number set than Jesus and the 12 disciples. We don't even need a long story for that.

What is interesting though, is what that 12 and 1

motif means for life.

Once you pass 12 still (dead) images per second, the brain creates motion. The dead are brung to life after 12! If you could not pack cells very tight together complex life could not exist.

Recap

#12 - kissing number how cells are arranged
Light is the unifying and organizing force
Jesus is the light of the World. All life is made with this light, all matter is made with this light. Science confirms the subatomic nature of all matter is Light.

There are 3 important kissing numbers that created life. There is a Hidden 1 with each motif... Even the Days are 6 and 1...

2 Adam & Eve & God/Snake - 3 total
6 Solomon Shield & Knowledge - 7 total
12 Disciples & Jesus - 13 total

Life and Light - 13 images per second is how your brain creates motion! Do you get that? Cells make tissues by using the 12 around 1 mold! After conception this is super obvious in how the Morula pronounced Moor Ruler is formed.

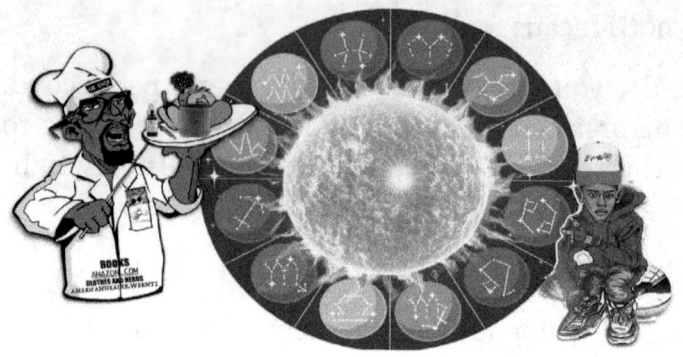

Matthew 6:10

Set the world right; Do what's best - as above, so below.

That's the message bible from the King James Version...

Thy kingdom come. Thy will be done in earth, as it is in heaven.

These translations can get crazy though... In this instance we don't have that problem.

Thoughts
as above, so below = in earth, as it is in heaven. The do what's best part is how they slip some Crowley/Blavatsky in the text... under the guise of Hermeticism. Hermeticism is Hermaphroditism wakey wakey... The devil is busy... Point being many young False Prophets have regurgitated Satanist telling you the Bible is fake because its really about astrology... They then go deep and show you how the sun declines on the southern cross below the northern horizon dec 22 - 25 3 days (72 hours)... 12 disciples = 12 zodiac signs etc...

Do you know they are actually proving the power and divinity of the Bible? Listen, the Bible is a math book of all numbers, which when turned into words tell intricately complex human stories about a particular bloodline, while describing the movements of the stars as well as? Vortex

Math comes from the first verse in Genesis, the 12 and 1 motif is how cells become tissues... Do you understand that no man or group of men have written a book in modern times with this many layers of complexity and coherence? Bwahahahaha...

How did the authors of the Bible know that lineage is based on men, not women? Women have a 28 day sun cycle not a moon cycle, that's the devil again. The moon doesn't have it's own light and doesn't reflect Infrared. Men have a 64 day cycle. I mean Terrence Howard is everywhere talking bout the flower of Life, he hasn't once said, the flowers of life and death bloom within.

SELF DESCRIPTIVE (REFERENCING) NUMBER

Number - c. 1300, "sum, **aggregate of a collection**," from Anglo-French noumbre, Old French nombre and directly from Latin numerus "a number, **quantity**," from PIE root *nem- "assign, allot; take."

The meaning "**written symbol or figure of arithmetic value**" is from late 14c. The meaning "single (numbered) issue of a magazine" is from 1795. The colloquial sense of "a person or thing" is by 1894. The meaning "dialing combination to reach a particular telephone receiver" is from 1879; hence wrong number (1886).

The sense of "**musical selection**" (1885) is from popular theater programs, where acts were marked by a number. Earlier numbers meant "**metrical sound or utterance, measured or**

harmonic expression" (late 15c.) and, from 1580s, "poetical measure, poetry, verse."

Number one "oneself" is from 1704 (mock-Italian form numero uno attested from 1973); the biblical Book of Numbers (c. 1400, Latin Numeri, Greek Arithmoi) is so called because it begins with a census of the Israelites. Childish slang number one and number two for "urination" and "defecation" attested from 1902. Number cruncher is 1966, of machines; 1971 of persons.

To get or have (someone's) number "have someone figured out" is attested from 1853; to say one's number is up (1806) meaning "one's time has come" is a reference to the numbers on a lottery, draft, etc. The numbers "illegal lottery" is from 1897, American English. Do a number on is by 1969, exact meaning unclear; by the early 1970s it can mean "emotionally manipulate" (1970), "damage or injure" (1975), or "assassinate, kill" (1971). The 1972 book of gay slang The Queen's Vernacular says it is synonymous with game, as well as with trick in the prostitution and magical senses, and defines it as "one's skit, act, schtick; contrived actions used to gain attention." The image may be of a routine song-and-dance performance, which if so makes it from the "**musical selection**" sense.

Pray - early 13c., preien, "**ask earnestly**, beg

(someone)," also (c. 1300) in a religious sense, "**pray to a god or saint**," from Old French preier "to pray" (c. 900, Modern French prier), from Vulgar Latin *precare (also source of Italian pregare), from Latin precari "ask earnestly, beg, entreat," from *prex (plural preces, genitive precis) "prayer, request, entreaty," from PIE root *prek- "to ask, request, entreat."

From early 14c. as "**to invite**." The deferential parenthetical expression I pray you, "please, if you will," attested from late 14c. (from c. 1300 as I pray thee), was contracted to pray in 16c. Related: Prayed; praying.
Praying mantis attested from 1809 (praying locust is from 1752; praying insect by 1816; see mantis). The Gardener's Monthly of July 1861 lists other names for it as camel cricket, soothsayer, and rear horse.

I know that prayer works via the waveguide, I know that it has to work for the sake of the planet. The scale though, Billions of people at once. Whose prayers do you answer? Which prayers do you answer? Does God ignore prayers? The best way for me to understand and relate what I know is to see the earth as a living being, a living cell. Humans are the enzymes in this cell. The cell requires new information (especially genetic diversity) as nutrition, however the foundations of the cell must be maintained. In a fast paced, ever changing environment the needs of the enzymes will constantly change. The cell needs its enzymes equipped to do their jobs. Prayer is the system built in to keep the enzymes equipped for their jobs. In L'Goat book we discuss how each individual cell in your body builds, breaks down, recycles and rebuilds all it's enzymes and proteins, done in earth as it is in heaven, right? Your heart, your intention, the requirement for your purpose, the exponential domino effect of your blessing, the way the response to your blessing affects your emotional state (and it's domino effect), all contribute to the response or lack there of. **<u>If the world don't need you, the world won't feed you</u>**. I think the problem is people are growing more and more selfish, praying to beat cases when they are guilty and haven't changed, praying for senseless material wealth at the expense of their peers, which is where Satanism comes in. I took a different route in life but I am from one of the original black hustler bloodlines in Harlem. I know the real essence of being a snitch is Satanism. Sacrifice the whole for the individual, as opposed to self sacrifice for the whole. That is the major take away that anyone who has ever heard of Jesus should have. The Leader gets the praise but he bears the weight of the challenge, that is what a cornerstone is.

Cornerstone - also *corner-stone*, late 13c., "**stone which lies at the corner of two walls** and <u>**unites them**</u>" (often the starting point of a building), hence, figuratively, "**that on which anything is founded**;" from **corner** (n.) + **stone** (n.). The figurative use is biblical (Isaiah xxvii.16, Job xxxviii.6, Ephesians ii.20), rendering Latin *lapis angularis*.

In U.S. history, Alexander H. Stephens's Cornerstone *speech* explaining **the new Confederate constitution was given at Savannah, Georgia, March 21, 1861**. The image is older in U.S. political discourse and originally referred to the federal union.

I endorse without reserve the much abused sentiment of Governor M'Duffie, that "**Slavery is the corner-stone of our republican edifice**;" while I repudiate, as ridiculously absurd, that much lauded but nowhere accredited dogma of Mr. Jefferson, that "**all men are born equal**." No society has ever yet existed, and I have already incidentally quoted the highest authority to show that none ever will exist, without a natural variety of classes. [James H. Hammond, "Letter to an English Abolitionist" 1845]

Keep in mind that while I of course detest slavery among humans, I do understand we have a live of servitude to God. I will go even further to state that is the basis for being on 'Santas Naughty or Nice list'... The purpose of the myth is early training for good adults, to be selfless and charitable without expectation of receiving. To whom much is given... much os required because

<u>WHAT YOU ARE GIVEN IS TO BE USED FOR THE WHOLE</u>. This can only be accomplished in my opinion in a capitalist system. Work is proof of character, character determines how we respond under pressure, pressure is what makes a piece of rough coal into a diamond.

Its a lot going on though, so you need to let God know what obstacles your facing, handling your business. Wait Doc, if God knows all, why pray? This is a mathemagickal system, Gods knows as in, is aware. God's body is information, you need to press the but on what to do with that information, stop/go, left/right, repeat/delete etc... You have free Will.

Will - Old English *willan, wyllan "**to wish, <u>desire</u>**; be willing; be used to; be about to" (past tense *wolde*), from Proto-Germanic *willjan (source also of Old Saxon *willian*, Old Norse *vilja*, Old Frisian *willa*, Dutch *willen*, Old High German *wellan*, German *wollen*, Gothic *wiljan* "to will, wish, desire," Gothic *waljan* "to choose").

The Germanic words are from PIE root *wel- (2) "to wish, will" (source also of Sanskrit vrnoti "chooses, prefers," varyah "**<u>to be chosen</u>**, eligible, excellent," varanam "**<u>choosing</u>**;" Avestan verenav- "to wish, will, choose;" Greek elpis "hope;" Latin volo, velle "to wish, will, desire;" Old Church Slavonic voljo, voliti "to will," veljo, veleti "to command;" Lithuanian velyti "to wish, **<u>favor</u>**," pa-velmi "I

will," viliuos "I hope;" Welsh gwell "better").
Compare also Old English wel "well," literally "according to one's wish;" wela "well-being, **riches**." The use as a future auxiliary was already developing in Old English. **The implication of intention or volition distinguishes it** from shall, which expresses or implies **obligation** or **necessity**. Contracted forms, especially after pronouns, began to appear 16c., as in sheele for "she will." In early use often -ile to preserve pronunciation. The form with an apostrophe ('ll) is from 17c.

This entire book may sound crazy or even delusional to you, the problem is, science agrees with me. If you discount anything that I have said, you have to find a plausible explanation for Quantum Entanglement & the Double Slit Experiment (the observer effect). THE FIRST THING WE HAVE TO DO IS SHOW YOU TRUTH FROM YOUR ENEMY. We have dealt with Jesus being a Naggar already, here I have to bring your attention to the term spook. Spook was just as or Moor disrespectful than the word N!gger. We began to use that term for the House Naga with the Field Naga spirit, the one who spied on 'Massa', he stole medicine, books and taught people how to read. We basically took that word, made it a slang, called each other spook with love and poof! It was quietly removed from usage though, why? Max Planck & Albert Einstein, they realized the 'black stuff' was the most information dense, and if there was a immaterial intelligence it would be found there. Fifty is a lot smarter than you think, Power is a modern adaptation of the Spook who sat by the door, the star is named Ghost!

Spook - 1801, "spectre, **apparition**, ghost;" first attested in a comical dialect poem, credited to "an old Dutch man in Albany" and printed in Vermont and Boston newspapers, which credited it to Springer's Weekly Oracle in New London, Conn.
The word is from Dutch spook, from Middle Dutch spooc, spoocke "a spook, a **ghost**," from a common Germanic source (German Spuk "ghost, apparition, hobgoblin," Middle Low German spok "spook," Swedish spok "scarecrow," Norwegian spjok "ghost, specter," Danish spøg "joke"), a word of unknown origin.

OED finds "No certain cognates." According to Klein's sources, possible outside connections include Lettish spigana "**dragon**, witch," spiganis "will o' the wisp," Lithuanian spingu, spingėti "**to shine**," Old Prussian spanksti "spark." Century Dictionary writes "There is nothing to show any connection with Ir. puca, **elf**, **sprite** ...."

The word also entered American English by 1830 as spuke, shpook, at first in the German-settled regions of Pennsylvania, via Pennsylvania Dutch Gschpuck, Schpuck, from German Spuk.

Meaning "superstition" is by 1918; as "**superstitious person**" perhaps by 1901. In 1890 a less refined **word for a spiritualist** or medium was spookist. Spooktacular, a pun on spectacular, is by 1945. Spook show meaning "frightening display" is by 1880, as "popular exhibition of legerdemain, **mentalism** or staged **necromancy**" by 1910. Spook house "abandoned house" is by 1857, as "haunted house" by 1866.

The sense "**Black person**" is attested by 1938, originally in **African-American slang** and not typically used with a sense of disparagement, nor considered an offensive word. **Black pilots** trained at Tuskegee Institute during World War II called themselves the Spookwaffe (a play on Luftwaffe): Col. [Richard S.] Harder (Rip to his friends for reasons known only to us of the old "Spookwaffe," as it was fondly known during World War II) has had a distinguished career in the Air Force and deserves mention along with many other fine officers who are a product of the formerly "**Negro Air Force**." [Lt. Col. George E. Franklin, USAF (Ret.), in Ebony, Oct. 1968]
The word was used earlier in aeronautic jargon to mean "novice pilot" (1939), which might also have influenced this particular use.

"Spooks" are a standard feature of every U.S. airport. They are the air-hungry amateur and novice pilots who haunt the hangars, hire planes, and learn to fly, not just to get places or save time but mostly for the fun of flying. [Life, Sep 11, 1939]

It is also attested as "a white jazz musician" by 1939, and as a disparaging term for a white person by 1947, possibly 1942, in the works of Nelson Algren (whose works also include the term used for **Black characters**). **The derogatory racial sense of "Black person" is attested from 1945, U.S., originally in hep-cat slang and defined specifically as "frightened negro" (compare spooky for sense development), used generally as a disparaging term for a Black person by 1953**. Green's Dictionary of Slang also proffers it as a slur for Italians and for Chinese/Vietnamese, though his examples might be attributable to other senses of the word.

The meaning "**undercover agent**" is attested from 1942. In student slang, a spook could be an unattractive girl (c. 1942), or a quiet, introverted student similar to a nerd (c. 1945).

1867, "walk or act like a ghost, **play the spook**," a sense now rare or obsolete, from spook (n.). The transitive meaning "frighten and unnerve" is from 1935; the intransitive meaning "become frightened" is by 1928. Related: Spooked; spooking.
also from 1867

Spooky - 1854, "frightening;" by 1889, "easily frightened," from spook (n. or v.) + -y (2).

Related: Spookily; **spookiness**. Alternative spookish is by 1847 (American English) as "like a ghost."

Spooky action(s) at a distance, a term used by Albert Einstein for what is now called "quantum entanglement," is by 1971, translating the original German spukhafte Fernwirkung.

Quantum - 1610s, "**sum, amount**," from Latin quantum (plural quanta) "**as much as**, so much as; **how much**? how far? how great an extent?" neuter singular of correlative pronominal adjective quantus "as much" (see quantity). **The word was introduced in physics directly from Latin by Max Planck**, 1900, on the notion of "**minimum amount of a quantity which can exist**;" reinforced by Einstein, 1905. Quantum theory is from 1912; quantum mechanics, 1922. The term quantum jump "**abrupt transition from one stationary state to another**" is recorded by 1954; quantum leap "sudden large advance" (1963), is often figurative.

Tangle - mid-14c., tanglen, "encumber, enmesh, **knit together confusedly**," a shortening of entangle in some cases, in others probably a nasalized variant of tagilen "**to involve** in a difficult situation, entangle," from a Scandinavian source (compare dialectal Swedish taggla "**to disorder**," Old Norse þongull "**seaweed**"), from Proto-Germanic *thangul- (source also of Frisian tung, Dutch tang, German Tang "**seaweed**").

If so, the original sense might be "**seaweed**" as something that entangles (itself, or oars, or fishes, or nets). "The development of such a verb from a noun of limited use like tangle 1 is somewhat remarkable, and needs confirmation" [Century Dictionary].

The transitive sense of "bring others into one's power, entrap" is from mid-15c. In reference to material things, from c. 1500. The meaning "fight with" is American English, recorded by 1928. Related: Tangled; tangling. Tanglefoot (1859) was Western U.S. slang for "strong whiskey." Tanglesome "complicated" is attested from 1823.

Entangle - early 15c., *entanglen*, "involve (someone in difficulty); embarrass;" from Anglo-French *entangler*, variant of *entagler*. See **en-** (1) + **tangle** (n.). Related: *Entangled*; *entangling*.

Entanglement - 1630s, "that which entangles," from entangle + -ment. From 1680s as "act of entangling." Foreign entanglements does not appear as such in Washington's Farewell Address (1796), though he warns against them. The phrase is found in William Coxe's 1798 memoirs of Sir Robert Walpole.

Quantum Entanglement - Spooky Action at a Distance

Observe - late 14c., observen, "to hold to (a manner of life or **course of** **conduct**), carry out the dictates of, attend to in practice, to keep, follow," from Old French observer, osserver "to observe, watch over, follow" (10c.), from Latin observare "watch over, note, heed, look to, attend to, guard, regard, comply with," from ob "in front of, before" (see ob-) + servare "to watch, keep safe," from PIE root *ser- (1) "to protect." Sense of "watch, perceive, notice" is from 1560s, via the notion of "see and note omens." Meaning "to say by way of remark" is from c. 1600. Related: Observed; observing.

Observer - 1550s, "one who keeps a rule, custom, etc.," agent noun from observe. Meaning "one who watches and takes notice" is from 1580s; this is the sense of the word in many newspaper names. Meaning "one who observes without participating" (at a meeting, conference, etc.) is by 1925.

Observance - mid-13c., observaunce, "act performed in accordance with prescribed usage," especially a religious or ceremonial one; late 14c., "care, concern, act of paying attention (to something)," from Old French observance, osservance "observance, discipline," and directly from Latin observantia "act of keeping customs, attention, respect, regard, reverence," from observantem (nominative observans), present participle of observare "watch over, note, heed, look to, attend to, guard, regard, comply with," from ob "in front of, before" (see ob-) + servare "to watch, keep safe," from PIE root *ser- (1) "to protect." Observance is the attending to and carrying out of a duty or rule. Observation is watching, noticing.

Observation - late 14c., observacioun, "the performance of a religious rite," from Old French observation (c. 1200) and directly from Latin observationem (nominative observatio) "a watching over, observance, investigation," noun of action from past-participle stem of observare "watch over, note, heed, look to, attend to, guard, regard, comply with," from ob "in front of, before" (see ob-) + servare "to watch, keep safe," from PIE root *ser- (1) "to protect." Sense of "act or fact of paying attention" is from 1550s. Meaning "a remark in reference to something observed" is recorded from 1590s.

Mote - "small particle, as of dust visible in a ray of sunlight," Old English mot, of unknown origin; perhaps related to Dutch mot "dust from turf, sawdust, grit," Norwegian mutt "speck, mote, splinter, chip." Hence, anything very small. Many references are to Matthew vii.3.

Particle - late 14c., "a bit or fragment, small

part or division of a whole, minute portion of matter," from Latin particula "little bit or part, grain, jot," diminutive of pars (genitive partis) "a part, piece, division" (from PIE root *pere- (2) "to grant, allot"). In grammar, "a part of speech considered of minor consequence or playing a subordinate part in the construction of a sentence" (1530s). Particle physics, which is concerned with sub-atomic particles, is attested from 1969. In construction, particle board (1957) is so called because it is made from chips and shavings of wood.

Wave - "move back and forth," Old English wafian "to wave, **fluctuate**" (related to wæfre "**wavering**, restless, **unstable**"), from Proto-Germanic *wab- (source also of Old Norse vafra "to hover about," Middle High German waben "to wave, **undulate**"), possibly from PIE root *(h)uebh- "**to move to and fro**; to weave" (see weave (v.)). Transitive sense is from mid-15c.; meaning "to make a sign by a wave of the hand" is from 1510s. Related: Waved; waving.
I was much further out than you thought
And not waving but drowning.
[Stevie Smith]

"moving billow of water," 1520s, alteration (by influence of wave (v.)) of Middle English waw, which is from Old English wagian "to move to and fro," from Proto-Germanic *wag- (source also of Old Saxon, Old High German wag, Old Frisian weg,

Old Norse vagr "water in motion, wave, billow," Gothic wegs "tempest"), probably from PIE root *wegh- "to go, move." The usual Old English word for "moving billow of water" was yð.

The "hand motion" meaning is recorded from 1680s; meaning "undulating line" is recorded from 1660s. Of people in masses, first recorded 1852; in physics, from 1832. Sense in heat wave is from 1843. The crowd stunt in stadiums is attested under this name from 1984, the thing itself said to have been done first Oct. 15, 1981, at the Yankees-A's AL championship series game in the Oakland Coliseum; soon picked up and popularized at University of Washington. To make waves "cause trouble" is attested from 1962.

Brain Wave - "**apparent telepathic vibration transferring a thought from one person to another without any other medium**," 1869, from brain (n.) + wave (n.).

Short Wave - in reference to **radio wavelength less than c.100 meters**, by 1907, from the noun phrase short wave, attested by 1839 in **electromagnetics**; see short (adj.) + wave (n.).

Be - Old English beon, beom, bion "be, **exist**, come to be, **become**, **happen**," from Proto-Germanic *biju- "**I am**, I will be." This "b-root" is from PIE root *bheue- "to be, exist, **grow**," and in addition to the words in English it yielded the German present first and second person singular (bin, bist, from Old High German bim "I am," bist "thou art"), the Latin perfective tenses of esse (fui "I was," etc.), Old Church Slavonic byti "be," Greek phu- "become," Old Irish bi'u "**I am**," Lithuanian būti "to be," Russian byt' "to be," etc.

The modern verb to be in its entirety represents the merger of two once-distinct verbs, the "b-root" represented by be and the am/was verb, which was itself a conglomerate. Roger Lass ("Old English") describes the verb as "a collection of semantically related paradigm fragments," while Weekley calls it "an accidental conglomeration from the different Old English dial[ect]s." **It is the most irregular verb in Modern English and the most common**. Collective in all Germanic languages, it has eight different forms in Modern English: BE (infinitive, subjunctive, imperative); AM (present 1st person singular); ARE (present 2nd person singular and all plural); IS (present 3rd person singular); WAS (past 1st and 3rd persons singular); WERE (past 2nd person singular, all plural; subjunctive); BEING (progressive & present participle; gerund); BEEN (perfect participle).

The paradigm in Old English was: eom, beo (present 1st person singular); eart, bist (present 2nd person singular); is, bið (present 3rd person singular); sind, sindon, beoð (present plural in all persons); wæs (past 1st and 3rd person singular);

wære (past 2nd person singular); wæron (past plural in all persons); wære (singular subjunctive preterit); wæren (plural subjunctive preterit).

The "b-root" had no past tense in Old English, but often served as future tense of am/was. In 13c. it took the place of the infinitive, participle and imperative forms of am/was. Later its plural forms (we beth, ye ben, they be) became standard in Middle English and it made inroads into the singular (I be, thou beest, he beth), but forms of are claimed this turf in the 1500s and replaced be in the plural. For the origin and evolution of the am/was branches of this tangle, see am and was.

That but this blow Might be the be all, and the end all. ["Macbeth" I.vii.5]

word-forming element of verbs and nouns from verbs, with a wide range of meaning: "about, around; thoroughly, completely; to make, cause, seem; to provide with; at, on, to, for;" from Old English be- "about, around, on all sides" (the unstressed form of bi "by;" see by (prep.)). The form has remained by- in stressed positions and in some more modern formations (bylaw, bygones, bystander).

The Old English prefix also was used to make transitive verbs and as a privative prefix (as in behead). The sense "on all sides, all about" naturally grew to include intensive uses (as in bespatter "**spatter** about," therefore "spatter very much," **besprinkle**, etc.). Be- also can be causative, or have just about any sense required. The prefix was productive 16c.-17c. in forming useful words, many of which have not survived, such as bethwack "to thrash soundly" (1550s) and betongue "to assail in speech, to scold" (1630s).

Effect - mid-14c., "execution or completion (of an act)," from Old French efet (13c., Modern French effet) "result, execution, completion, ending," from Latin effectus "accomplishment, **performance**," from past participle stem of efficere "**work out**, accomplish," from assimilated form of ex "out" (see ex-) + combining form of facere "to make, to do" (from PIE root *dhe- "to set, put"). From French, borrowed into Dutch, German, Scandinavian.

From late 14c. as "power or capacity to produce an intended result; efficacy, effectiveness," and in astrology, "operation or action (of a heavenly body) on human affairs; **influence**." Also "that which follows from something else; a consequence, a result." From early 15c. as "intended result, purpose, object, **intent**." Also formerly with a sense of "reality, fact," hence in effect (late 14c.), originally "in fact, actually, really." Meaning "**impression produced on the beholder**" is from 1736. Sense in stage effect, sound effect, etc. first recorded 1881.

"**to produce as a result**; to bring to a desired end,"

1580s, from Latin effectus, past participle of efficere "work out, accomplish" (see effect (n.)). Related: Effecting; effection; effectible.

Ob - word-forming element meaning "**toward**; **against**; before; near; across; down," also used as an intensive, from Latin ob (prep.) "**in the direction of**, in front of, before; toward, to, at, upon, about; in the way of; with regard to, because of," from PIE root *epi, also *opi "near, against" (see epi-).

Server - late 14c., "**one who serves**" in any capacity, agent noun from serve (v.). Especially "**an attendant at a meal**" (mid-15c.). By 1580s in sports. The meaning "that which is used in serving" is by c. 1600; **the computing sense is by 1992**.

Quantum is Math, Arithmetic (cause God is Music)!

Observe is in the Bible over 50 times bwahahahaha... Observer = Influencer did you know that? The Devil keeps you operating in the spirit but blind so you serve him...

The Observer Effect - The observer effect is **the fact that observing a situation or phenomenon necessarily changes it**. Observer effects are especially prominent in physics where observation and **uncertainty** are fundamental aspects of modern quantum mechanics. In physics, the observer effect is the disturbance of an observed system by the act of observation. This is often the result of utilising instruments that, by necessity, alter the state of what they measure in some manner. Ask yourself now why scientist actually keep trying to observe moor and moor of the universe? Candace Owens said it best 'What the hell is NASA actually doing"?

John Whiteside Parsons (born Marvel Whiteside Parsons; October 2, 1914 – June 17, 1952) was an American rocket engineer, chemist, and Thelemite occultist. Parsons was one of the principal founders of both the Jet Propulsion Laboratory (JPL) and the Aerojet Engineering Corporation. He invented the first rocket engine to use a castable, composite rocket propellant, and pioneered the advancement of both liquid-fuel and solid-fuel rockets. In 1939, Parsons converted to Thelema, a religious movement founded by English occultist Aleister Crowley. Parsons and his first wife, Helen Northrup, joined the Ordo Templi Orientis (OTO); he became the California OTO branch leader in 1942. Historians of Western esotericism cite him as a prominent figure in propagating Thelema in North America. Parsons was dismissed from JPL and Aerojet in 1944, due to his involvement with OTO and his hazardous laboratory practices. In 1945, he and his wife divorced. In 1946, he married Marjorie Cameron. Shortly afterward, L.

Ron Hubbard defrauded Parsons of his life savings. Lafayette Ronald Hubbard (March 13, 1911 – January 24, 1986) was an American author and the founder of Scientology. Didn't Rizzo Islam family merge the NOI with Scientology? Allegedly, bwahahahaha..... Yall better get right with the Lord, periodT.

Anyway....

You may not have known it, the birth of AI began with the Double Slit Experiment. You guys have to watch these people, "**the need for the "observer" to be conscious is not supported by scientific research**, and has been pointed out as a misconception rooted in a poor understanding of the quantum wave function ψ and the quantum measurement process".

Not only is this the root of AI but also many satanic rituals, Rick Ross any many other Satanists are so comfortable they tell you if you'd listen... "put Molly all in her champagne, she aint even know it. I took her home and I enjoyed that, and she aint even know it". You see we, well you guys assume that when they knock people out its so they can have sex with people. Yall are crazy if you think Millionaires and Billionaires especially sexy or handsome ones need to knock people out for sex... You have no clue what these demons are doing... Just know they are all moving under Crowley & Blavatsky, their 'gateway drug' is Chakras, the catch phrase is anal chakra. In fact there are a few phrases:

Anal Chakra - the key to spirituality

If you don't think this is all Satanic, then why use God's Covenant? The Rainbow is God's covenant with man you know... Your now running from the rainbow like a Vampire from a Cross... The Rainbow and the Cross represent the Trinity, DO NOT BE FOOLED!

Oooohhh Wait this is good one, Jesus already paid for our sins, and no sin is worse than another, under these precepts all manor of Hell may be raised.
Why not rob, steal, kill, rape go wild... This is why the cast of the View is so opposed to children being taught the 10 commandments but in favor of these very same students being taught TransCulture! How Swaye? How are the children to young and innocent to learn the Bible but not
_____... Bruh! If you can't see now what going on, you will never wake up!
Being a 'sinner' isn't about loving sin and being full aware of your sin, with all intent on continuing to sin, even convincing others that sin isn't sin. Bwahahahaha this is like living in a scary movie! You not only need to pray with your Rosary, you need to wear it daily!

Dr. York taught you that the cross was the sign of death and the mark of the Beast while raping over 40 kids! He taught his students to rape kids like Natureboy and Polight, his son Jacob York when into the music business with Puff Daddy (allegedly). Are you Nagas awake? Natureboy had a silent partner (allegedly) though his name was Miguel or something, a college professor like Jabari, like Marc Lamont Hill... He funded Natureboy through a barbershop at first, then throughout his entire campaign, even when he confessed on YouTube to letting children play with his ____ post ejaculation. These men like Jabari are still out there, Marc Lamont Hill on TV!!! Yall better watch out!!! Stay prayed up time, the time to learn in peace is over, you gotta put that spiritual armor on now!

Leviticus 18:22 ~ You shall not lie with a male as with a woman; it is an abomination.

Leviticus 20:13 ~ If a man lies with a male as with a woman, both of them have committed an abomination; they shall surely be put to death; their blood is upon them.

Romans 1:26-28 ~ For this reason God gave them up to dishonorable passions. For their women exchanged natural relations for those that are contrary to nature; and the men likewise gave up natural relations with women and were consumed with passion for one another, men committing shameless acts with men and receiving in themselves the due penalty for their error. And since they did not see fit to acknowledge God, God gave them up to a debased mind to do what ought not to be done.

1 Corinthians 6:9-10 - Or do you not know that the unrighteous will not inherit the kingdom of God? Do not be deceived: neither the sexually immoral, nor idolaters, nor adulterers, nor men who practice homosexuality, nor thieves, nor the greedy, nor drunkards, nor revilers, nor swindlers will inherit the kingdom of God.

Again the same fashion as anger and violence causes Heart Disease, Male on Male anal sex causes Colon Cancer, AIDs etc... These punishments are happening in real time, in real life... Don't get tricked out of position!

Anyway... Oooohhh Wait this is good one, Jesus already paid for our sins, and no sin is worse than another. That was the Democrat war cry under Kabala Harris (shout out AdeptHNIC)!!! Here is what you must understand, that is the foundation behind 'Do as though whilst/whilt'... All these folks twisted!

You have Capacitor Hands, to build Tribo electricity, the more voltage the more powerful the prayer etc.. Get busy, get you a gold, copper, bronze, brass, stainless steel (metal or wood beads) Rosary! Obviously that all depends on budget... those are options, right now 6 real gold rosaries a little outside my budget but... stuff like that in 2024 holiday season... Books, Prayer Beads, Herbs... Lock in family cause the Healthcare System is about to go for a ride... You need to be ready!!! You need to stockpile your stuff now... AmericanHealer.Website don't say I didn't warn you! Don't let these people tell you, you can receive blessing without Faith.

Faith - mid-13c., faith, feith, fei, fai "faithfulness to a trust or promise; **loyalty to a person**; honesty, truthfulness," from Anglo-French and Old French feid, foi "faith, belief, trust, **confidence**; pledge" (11c.), from Latin fides "trust, faith, **confidence**, reliance, credence, belief," from root of fidere "to trust,"from PIE root *bheidh- "to trust, confide, persuade." For sense evolution, compare belief. It has been accommodated to other English abstract nouns in -th (truth, health, etc.).

From early 14c. as "assent of the mind to the truth of a statement for which there is incomplete evidence," especially "**belief in religious matters**" (matched with hope and charity). Since mid-14c. in reference to the Christian church or religion; from late 14c. in reference to any religious persuasion.

And faith is neither the submission of the reason, nor is it the acceptance, simply and absolutely upon testimony, of what reason

cannot reach. Faith is: the being able to cleave to a power of goodness appealing to our higher and real self, not to our lower and apparent self. [Matthew Arnold, "Literature & Dogma," 1873]

From late 14c. as "**confidence in a person or thing with reference to truthfulness or reliability**," also "fidelity of one spouse to another." Also in Middle English "a sworn oath," hence its frequent use in Middle English oaths and asseverations (par ma fay, mid-13c.; bi my fay, c. 1300).

Faith means confidence, once you pray on it, for it and around it, you must approach it! You have to walk in Faith (not ignorance), walk confidently knowing that 2+2=4. If you have $5 and you are walking to the store to make a $2 purchase, you have no doubt you will receive what you need from the store. Even with the $5 though, you have to get to the store, go to the isle, stand in line... That's why faith without works is dead, then of course you must embrace your responsibility.

Remember what your breath is, what's in your breath, don't make senseless, meaningless prayers, make focused prayers...

Spontaneous chemiluminescence of human breath. Spectrum, lifetime, temporal distribution, and correlation with peroxide

M D Williams, B Chance

Human breath spontaneously emits photons at a rate of approximately 7,000/liter-s. The emission has a peak in the red part of the spectrum and an ultraviolet contribution. The emission count rate correlates with peroxide concentration in a saturating manner under normal breathing conditions. When trapped in a balloon, the breath luminescence count rate has a half-decay time of approximately 20 min and exhibits more than one mode of decay. The photomultiplier pulses generated by breath luminescence arrive in bursts. The chemiluminescence process appears by these criteria to include chain reactions, long-lived emitters, or both.

Matthew 15

Then came to Jesus scribes and Pharisees, which were of Jerusalem, saying,
2 Why do thy disciples transgress the tradition of the elders? for they wash not their hands when they eat bread.
3 But he answered and said unto them, Why do ye also transgress the commandment of God by your tradition?
4 For God commanded, saying, Honour thy father and mother: and, He that curseth father or mother, let him die the death.
5 But ye say, Whosoever shall say to his father or his mother, It is a gift, by whatsoever thou mightest be profited by me;

6 And honour not his father or his mother, he shall be free. Thus have ye made the commandment of God of none effect by your tradition.

7 Ye hypocrites, well did Esaias prophesy of you, saying,

8 This people draweth nigh unto me with their mouth, and honoureth me with their lips; but their heart is far from me.

9 But in vain they do worship me, teaching for doctrines the commandments of men.

10 And he called the multitude, and said unto them, Hear, and understand:

11 Not that which goeth into the mouth defileth a man; but that which cometh out of the mouth, this defileth a man.

12 Then came his disciples, and said unto him, Knowest thou that the Pharisees were offended, after they heard this saying?

13 But he answered and said, Every plant, which my heavenly Father hath not planted, shall be rooted up.

14 Let them alone: they be blind leaders of the blind. And if the blind lead the blind, both shall fall into the ditch.

15 Then answered Peter and said unto him, Declare unto us this parable.

16 And Jesus said, Are ye also yet without understanding?

17 Do not ye yet understand, that whatsoever entereth in at the mouth goeth into the belly, and is cast out into the draught?

18 But those things which proceed out of the mouth come forth from the heart; and they defile the man.

19 For out of the heart proceed evil thoughts, murders, adulteries, fornications, thefts, false witness, blasphemies:

20 These are the things which defile a man: but to eat with unwashen hands defileth not a man.

21 Then Jesus went thence, and departed into the coasts of Tyre and Sidon.

22 And, behold, a woman of Canaan came out of the same coasts, and cried unto him, saying, Have mercy on me, O Lord, thou son of David; my daughter is grievously vexed with a devil.

23 But he answered her not a word. And his disciples came and besought him, saying, Send her away; for she crieth after us.

24 But he answered and said, I am not sent but unto the lost sheep of the house of Israel.

25 Then came she and worshipped him, saying, Lord, help me.

26 But he answered and said, **It is not meet to take the children's bread, and to cast it to dogs**.

27 And she said, Truth, Lord: **yet the dogs eat of the crumbs which fall from their masters' table**.

28 Then Jesus answered and said unto her, O woman, **great is thy faith**: be it unto thee even as thou wilt. And **her daughter was made whole from that very hour**.

29 And Jesus departed from thence, and came nigh unto the sea of Galilee; and went up into a

mountain, and sat down there.

30 And great multitudes came unto him, having with them those that were lame, blind, dumb, maimed, and many others, and cast them down at Jesus' feet; and he healed them:

31 Insomuch that the multitude wondered, when they saw the dumb to speak, the maimed to be whole, the lame to walk, and the blind to see: and they glorified the God of Israel.

32 Then Jesus called his disciples unto him, and said, I have compassion on the multitude, because they continue with me now three days, and have nothing to eat: and I will not send them away fasting, lest they faint in the way.

33 And his disciples say unto him, Whence should we have so much bread in the wilderness, as to fill so great a multitude?

34 And Jesus saith unto them, How many loaves have ye? And they said, Seven, and a few little fishes.

35 And he commanded the multitude to sit down on the ground.

36 And he took the seven loaves and the fishes, and gave thanks, and brake them, and gave to his disciples, and the disciples to the multitude.

37 And they did all eat, and were filled: and they took up of the broken meat that was left seven baskets full.

38 And they that did eat were four thousand men, beside women and children.

39 And he sent away the multitude, and took ship,

and came into the coasts of Magdala.

Jesus basically said listen Sis... Your daughter is healed because **YOU PUT THAT WORK IN**! At this point I can't even refer you to L'Goat book, let me just give you this, from that book:

You Ready?

Here are some 120 appetizers before we jump into the Water & Wine as the main course...

120 is equal to 5! (5 factorial) or 1 x 2 x 3 x 4 x 5.

Interestingly, the number 120 is equal to adding the first fifteen digits or 1 + 2 + 3 + 4 + 5 + 6 + 7 + 8 + 9 + 10 + 11 + 12 + 13 + 14 + 15.

Genesis **6:3** And the Lord said, My spirit shall not always strive with man, for that he also is flesh: yet his days shall be an hundred and twenty years.

Acts **1:15** At a time when about **120** disciples had gathered together, Peter got up and spoke to them.

1Kings **9:14** And Hiram sent to the king sixscore (**6x20**) talents of gold. Now how many gallons of Water did Jesus turn into Wine?

The same ATPase pump, with the F0 wall clock, that makes three **120** degree turns in humans, also does this in Plants. These **120** degree turns are powered by Water remember? Photosynthesis is literally the process of Hydrating the Magical Black Stuff, Carbon.

The etymology of Glucose is Wine (the first sugar formed on Earth as a crop, is Grape Sugar)! Osiris is said to have healed the world with Grapes and the

Waters of the Nile! The first fruit crop ever!! One of the first fruits period (maybe Dates or Bananas) but definitely the first Crop! Grapes have been used against people too... Slavery & Wine are old friends!

"In the daylight, photosynthesis is dominant, so there is a net release of oxygen. At night, photosynthesis stops but respiration continues, so there is a net consumption of oxygen."

Yes now you get it! Horus turning Water into Wine is about the rising Sun, turning water into HYDRATED-CARBON aka Carbohydrates!!!

Oops I meant Jesus performing Photosynthesis, wait oops I mean the Sun turning Water into Wine, hold up...

I meant Jesus, hold on.... Wasn't no J back then....
I meant to say Gases turning Water into Fruit. LMAO....

My apologies...

I want to end with this in case it wasn't clear, godless science is Satanism. The spoke word is how the world is created, in order to make a people or land you can just speak it into existence. Godless Science makes people speak with conviction, the conviction of their soul, speaking all manor of wickedness into existence, conjuring demons, practicing death rituals as holidays and everything else.... Stay prayed up!

The EnD? nO suCh ThInG!

Letter Name	Lower Case	Upper Case	Transliteration	Pronunciation and Notes
alpha	α	A	a	long open sound as in "father" (not "hay")
beta	β	B	b	
gamma	γ	Γ	g	γγ (double gamma) is pronounced "ng" as in "angle"
delta	δ	Δ	d	
epsilon	ε	E	e	short sound as in "bet" or "end"
zęta	ζ	Z	z	
ęta	η	H	ē	long open sound as in "hey" (not "feed") written with a <u>bar</u> over the letter in transliteration
thęta	θ	Θ	th	throat-less sound as in "thin" (not "this")
iota	ι	I	i	short sound as in "sit" (not "white") written *without a dot* over the Greek iota
kappa	κ	K	k	
lambda	λ	Λ	l	
mu	μ	M	m	
nu	ν	N	n	

xi	ξ	Ξ	ks	combined "k" and "s" sounds, as in "fox"
omicron	ο	Ο	o	short sound as in "box" or "off"
pi	π	Π	p	
rho	ϱ	Ρ	r	don't confuse this with the "pi"
sigma	σ, ς	Σ	s	regular σ at the start or middle of words; terminal ς is used at the end of words
tau	τ	Τ	t	
upsilon	υ	Υ	u (sometimes v or y)	short sound halfway between "put" and "pit" [more like French "tu" or German "
phi	φ	Φ	ph	like the regular "f" sound in "photo"
chi	χ	Χ	ch	breathy sound as in "Bach" or "chasm" (not "church")
psi	ψ	Ψ	ps	combined "p" and "s" sounds
omega	ω	Ω	ō	long sound as in "go" written with a bar over the letter in transliteration

NEXT

Greek is something I am newly interested in, I don't know what I will do with it but... Its there just in case you wanna play with it as well.

Particulars

Matthew 6:5-8

5 And when thou prayest, thou shalt not be as the hypocrites are: for they love to pray standing in the synagogues and in the corners of the streets, that they may be seen of men. Verily I say unto you, They have their reward.
6 But thou, when thou prayest, enter into thy closet, and when thou hast shut thy door, pray to thy Father which is in secret; and thy Father which seeth in secret shall reward thee openly.
7 But when ye pray, use not vain repetitions, as the heathen do: for they think that they shall be heard for their much speaking.
8 Be not ye therefore like unto them: for your Father knoweth what things ye have need of, before ye ask him.
Hebrews 4:16
16 Let us therefore come boldly unto the throne of

grace, that we may obtain mercy, and find grace to help in time of need.

1 Thessalonians 5:16–18

16 Rejoice evermore.

17 Pray without ceasing.

18 In every thing give thanks: for this is the will of God in Christ Jesus concerning you.

Philippians 4:6–7

6 Be careful for nothing; but in every thing by prayer and supplication with thanksgiving let your requests be made known unto God.

7 And the peace of God, which passeth all understanding, shall keep your hearts and minds through Christ Jesus.

1 John 5:14–15

14 And this is the confidence that we have in him, that, if we ask any thing according to his will, he heareth us:

15 And if we know that he hear us, whatsoever we ask, we know that we have the petitions that we desired of him.

Matthew 6:9–13

9 After this manner therefore pray ye: Our Father which art in heaven, Hallowed be thy name.

10 Thy kingdom come, Thy will be done in earth, as it is in heaven.

11 Give us this day our daily bread.

12 And forgive us our debts, as we forgive our debtors.

13 And lead us not into temptation, but deliver us

from evil: For thine is the kingdom, and the power, and the glory, for ever. Amen.

Mark 11:24

24 Therefore I say unto you, What things soever ye desire, when ye pray, believe that ye receive them, and ye shall have them.

Ephesians 6:17–18

17 And take the helmet of salvation, and the sword of the Spirit, which is the word of God:

18 Praying always with all prayer and supplication in the Spirit, and watching thereunto with all perseverance and supplication for all saints;

James 5:16

16 Confess your faults one to another, and pray one for another, that ye may be healed. The effectual fervent prayer of a righteous man availeth much.

This next verse hit hard, my lil brother's death brung me to knees, to call on the name of the Lord. I had no idea that the first 'prayers' in the Bible was in wake of murder.

genesis 4

8 And Cain talked with Abel his brother: and it came to pass, when they were in the field, that Cain rose up against Abel his brother, and slew him.

9 And the Lord said unto Cain, Where is Abel thy brother? And he said, I know not: Am I my brother's keeper?

26 And to Seth, to him also there was born a son;

and he called his name Enos: **then began men to call upon the name of the Lord**.

What is the purpose of calling upon the name of the Lord. *The Catholic Encyclopedia* says" the raising of the mind and soul to God". Your soul is forever entangled with it's source, the breath of life. Everyone is jumping on the Quantum Christianity train now, the Devil will be at the next stop be careful. Is anyone explaining how or why? Fractal Symmetry, it's only something noticeable from outside the system. Inside or apart of a fractal system, you can only figure it out by 'spooky action at a distance' type shift.

Fractal - "never-ending pattern," 1975, from French fractal, ultimately from Latin fractus "interrupted, **irregular**," literally "broken," past participle of frangere "to break" (from PIE root *bhreg- "to break"). Coined by French mathematician Benoit Mandelbrot (1924-2010) in "Les Objets Fractals."

Many important spatial patterns of Nature are either irregular or fragmented to such an extreme degree that ... classical geometry ... is hardly of any help in describing their form. ... I hope to show that it is possible in many cases to remedy this absence of geometric representation by using a family of shapes I propose to call fractals — or fractal sets. [Mandelbrot, "Fractals," 1977]

The term was suggested earlier in Mandelbrot's 1967 book, "How Long is the Coast of Britain -- Statistical Self-Similarity and Fractional Dimension."

If you go back, remember I was saying the disconnect between science and spirituality is 'duplication'. In order for a thing to be classified as real science, it must be duplicatable, the results or objects must be mass producible. I said that does not work with Life, each life is unique! The proof is always in Math and Geometry right? The math is irregular numbers rule the universe, pi, phi etc... The Geometry is Fractals, fractals are like solid holograms.

I just gave you the original definition of Fractal, let me give you the modern definition.

Fractal - a type of mathematical shape that are infinitely complex. In essence, a Fractal is a pattern that repeats forever, and every part of the Fractal, regardless of how zoomed in, or zoomed out you are, it looks very similar to the whole image; a never-ending pattern. Fractals are infinitely complex patterns that are self-similar across different scales.

Now that you have that, let me show you something, I have been saying until I have been blue in the face. Life is built on the Russian doll motif, we are exactly in the image and likeness of our creator, the problem is we KEEP LOOKING IN THE MIRROR! It is our cells, that are in the image and likeness of our creator, I showed you the breath of life and now I will show you the Fractal or Russian doll motif (again). I am hoping that every time we go over this it gets easier to understand.

Genesis 1:27

Does God have Palilalia or Tourettes? Why does Genesis 1:27 seem like God saying the same thing over and over?

Genesis 1:27

So God created man in his own image, in the image of God created he him; male and female created he them.

If you are not paying attention, nah... Pastors, Ministers and all other manor of bible folks have missed this!

So God created man in his own image - Man is inside of God!

in the image of God created he him - Man is smaller copy of God!

male and female created he them - God & Man are Binary.
I know they are going to say, you taking it

too literal Doc, I will absorb the criticism. The other side of the argument takes away intelligence from God. Why would god be repeating himself? I am saying in no uncertain terms that Genesis 1:27 describes a Fractal system. Einstein couldn't explain spooky action at a distance, I can (with the help of the G-Homey). When you exercise free will, you collapse the wave function of infinite probability into a set sequence of building energy states into matter or matters (pun intended).

Let's look at this a different way in the book 'the Rise of the Black Culture Vultures, we discuss how plants in a garden, park or forest form a coherent matrix, a body. People are no different, in a church, concert etc...

Concert - 1660s, "**agreement of two or more in design or plan**; **accord**, **harmony**," from French concert (16c.), from Italian concerto "concert, harmony," from concertare "bring into agreement," apparently from Latin concertare "to contend with zealously, contest, dispute, debate" from assimilated form of com "with" (see con-) + certare "to contend, strive," frequentative of certus, variant past participle of cernere "separate, distinguish, decide" (from PIE root *krei- "to sieve," thus "discriminate, distinguish").

The proposed sense evolution between Latin

("to contend with") and medieval Italian ("bring into agreement") seems extreme and is difficult to explain. Perhaps the shift is from "to strive against" to "to strive alongside" (compare English fight with), or perhaps it is via the notion of "confer, arrange by conference, debate for the sake of agreement." Some have suggested the sense shifted through confusion of Latin concertus with consertus, past participle of concerere "to join, fit, unite."

Sense of "public musical performance," usually of a series of separate pieces, is from 1680s, from Italian (Klein suggests Latin concentare "to sing together," from con- + cantare "to sing," as the source of the Italian word in the musical sense). The general sense of "any harmonious agreement or orderly union" is from 1796. Concert-master "first violinist of an orchestra" is from 1815, translating German Konzertmeister.

God is Music!

Matthew 18

At the same time came the disciples unto Jesus, saying, Who is the greatest in the kingdom of heaven?

2 And Jesus called a little child unto him, and set him in the midst of them,

3 And said, Verily I say unto you, Except ye be

converted, and become as little children, ye shall not enter into the kingdom of heaven.

4 Whosoever therefore shall humble himself as this little child, the same is greatest in the kingdom of heaven.

5 And whoso shall receive one such little child in my name receiveth me.

6 But whoso shall offend one of these little ones which believe in me, it were better for him that a millstone were hanged about his neck, and that he were drowned in the depth of the sea.

7 Woe unto the world because of offences! for it must needs be that offences come; but woe to that man by whom the offence cometh!

8 Wherefore if thy hand or thy foot offend thee, cut them off, and cast them from thee: it is better for thee to enter into life halt or maimed, rather than having two hands or two feet to be cast into everlasting fire.

9 And if thine eye offend thee, pluck it out, and cast it from thee: it is better for thee to enter into life with one eye, rather than having two eyes to be cast into hell fire.

10 Take heed that ye despise not one of these little ones; for I say unto you, That in heaven their angels do always behold the face of my Father

which is in heaven.

11 For the Son of man is come to save that which was lost.

12 How think ye? if a man have an hundred sheep, and one of them be gone astray, doth he not leave the ninety and nine, and goeth into the mountains, and seeketh that which is gone astray?

13 And if so be that he find it, verily I say unto you, he rejoiceth more of that sheep, than of the ninety and nine which went not astray.

14 Even so it is not the will of your Father which is in heaven, that one of these little ones should perish.

15 Moreover if thy brother shall trespass against thee, go and tell him his fault between thee and him alone: if he shall hear thee, thou hast gained thy brother.

16 But if he will not hear thee, then take with thee one or two more, that in the mouth of two or three witnesses every word may be established.

17 And if he shall neglect to hear them, tell it unto the church: but if he neglect to hear the church, let him be unto thee as an heathen man and a publican.

18 Verily I say unto you, **Whatsoever ye shall**

bind on earth shall be bound in heaven: and whatsoever ye shall loose on earth shall be loosed in heaven.

19 Again I say unto you, **That if two of you shall agree on earth as touching any thing that they shall ask, it shall be done for them of my Father which is in heaven**.

20 **For where two or three are gathered together in my name, there am I in the midst of them**.

21 Then came Peter to him, and said, Lord, how oft shall my brother sin against me, and I forgive him? till seven times?

22 Jesus saith unto him, I say not unto thee, Until seven times: but, Until seventy times seven.

23 Therefore is the kingdom of heaven likened unto a certain king, which would take account of his servants.

24 And when he had begun to reckon, one was brought unto him, which owed him ten thousand talents.

25 But forasmuch as he had not to pay, his lord commanded him to be sold, and his wife, and children, and all that he had, and payment to be made.

26 The servant therefore fell down, and

worshipped him, saying, Lord, have patience with me, and I will pay thee all.

27 Then the lord of that servant was moved with compassion, and loosed him, and forgave him the debt.

28 But the same servant went out, and found one of his fellowservants, which owed him an hundred pence: and he laid hands on him, and took him by the throat, saying, Pay me that thou owest.

29 And his fellowservant fell down at his feet, and besought him, saying, Have patience with me, and I will pay thee all.

30 And he would not: but went and cast him into prison, till he should pay the debt.

31 So when his fellowservants saw what was done, they were very sorry, and came and told unto their lord all that was done.

32 Then his lord, after that he had called him, said unto him, O thou wicked servant, I forgave thee all that debt, because thou desiredst me:

33 Shouldest not thou also have had compassion on thy fellowservant, even as I had pity on thee?

34 And his lord was wroth, and delivered him to the tormentors, till he should pay all that was due unto him.

35 So likewise shall my heavenly Father do also unto you, if ye from your hearts forgive not every one his brother their trespasses.

I can't hold you, this chapter is crazy! This chapter is 18, you know I be on them numbers, but the bars are crazy!

The cells in your body demonstrate Quantum Entanglement 24/7! Its just taken for granted because a) we understand that our cells are inside a system & b) most people don't understand the scale and complexity of the human body. It's there, Ill let you unpack it, from a simple circuit or embrace to a fractal system...

Let me give it to you one more way, this observer thing go crazy. Geometry determines how energy moves in a system. A fractal system is a bunch of identical shapes, they are just different sizes, and they are all connected. Energy pattern is based on th geometry of it's container...

Energy pattern creates a specific frequency, God is Music! The same notes in different octaves are **harmonically** related: **a harmonic series based on a low "C" note contains the frequencies of every higher C**. As such, these notes share a unique mathematical relationship with each other that they don't share with other notes. Notes are classified by their name and match up to a particular frequency (Hz) that portrays the number of vibrations per second. For example, 1 Hz = 1 vibration per second. In a fractal system, the parts of the whole are all humming the same tune, its a **UNIVERSE**.

GOD IS MUSIC!

Light travels around the earth at 7 times a second, the real circle 7.

Lets get crazier, did you know that numbers are the only thing that can explain Energy or Music? Do you know why? All of them are figments of your imagi**nation**. Image Nation is the world of number!

Imagination - "faculty of the mind which forms and manipulates images," mid-14c., ymaginacion, from Old French imaginacion "concept, mental picture; hallucination," from Latin imaginationem (nominative imaginatio) "imagination, a fancy," noun of action from past participle stem of imaginari "to form an image of, represent"), from imago "an image, a likeness," from stem of imitari "to copy, imitate" (from PIE root *aim- "to copy")

We discuss vortex math in this book, vortex math is based on the number 9. Have you ever seen a number 9 in real life? You may be thinking duh... 9 is right there but 9 is the 'Arabic Numeral' that represents the number 9, there are plenty of different 'numerals' that represent the number 9 ie... VIIII, IX, Θ´, Ո (tet) … see what I mean? In ancient egypt they would've used totally foreign representation for what we call the number 9. We have never actually seen and numbers! Number like light and sound, **ONLY EXIST IN OUR MIND**. Have you even processed the chicken or the egg concept? Better still, I told you Jesus represented the number 13, is that in a abstract or literal sense?

Is there a such thing a 'things moving' or is that only in our mind? Your brain has to exceed the frame rate to create movement. You need to see something at more than 12 times per second for it to begin 'moving'.

Frame - Old English framian "to profit, be helpful, avail, benefit," from fram (adj., adv.) "active, vigorous, bold," originally "going forward," from fram (prep.) "forward; from" (see from). Influenced by related Old English fremman "help forward, promote; do, perform, make, accomplish," and Old Norse fremja "to further, execute." Compare German frommen "avail, profit, benefit, be of use."
Sense focused in Middle English from "make ready" (mid-13c.) to "prepare timber for building" (late 14c.). Meaning "compose, devise" is first attested 1540s. The criminal slang sense of "blame an innocent person" (1920s) is probably from earlier sense of "plot in secret" (1900), perhaps ultimately from meaning "fabricate a

story with evil intent," which is first attested 1510s.

c. 1200, "profit, benefit, advancement;" mid-13c. "a structure composed according to a plan," from frame (v.) and in part from Scandinavian cognates (Old Norse frami "advancement"). In late 14c. it also meant "the rack."
Meaning "sustaining parts of a structure fitted together" is from c. 1400. Meaning "enclosing border" of any kind is from c. 1600; specifically "border or case for a picture or pane of glass" from 1660s. The meaning "human body" is from 1590s. Of bicycles, from 1871; of motor cars, from 1900. Meaning "separate picture in a series from a film" is from 1916. From 1660s in the meaning "particular state" (as in Frame of mind, 1711). Frame of reference is 1897, from mechanics and graphing; the figurative sense is attested from 1924.

Rate - early 15c., "estimated value or worth, proportional estimation according to some standard; monetary amount; a proportional part," from Old French rate "price, value" and directly from Medieval Latin rata (pars) "fixed (amount)," from Latin rata "fixed, settled," fem. past participle of reri "to reckon, think" (from PIE root *re- "to reason, count").
Meaning "degree of speed" (properly ratio between distance and time) is attested from 1650s. Currency exchange sense of "basis of equivalence upon which one currency is exchanged for

another" is recorded by 1727. Meaning "fixed public tax assessed on property for some local purpose" is by 1712.

First-rate, second-rate, etc. are 1640s, from British Navy division of ships into six classes based on size and strength. Phrase at any rate originally (1610s) meant "at any cost," hence "positively, assuredly." weakened sense of "at least" is attested by 1760. Rate-payer "one who is assessed and pays a local tax" is by 1825.

"to scold, chide vehemently, rebuke," late 14c., raten, probably from Old French rateir, variant of reter "to impute blame, accuse, find fault with," from Latin reputare "to count over, reflect," in Vulgar Latin, "to impute, blame," from re- "repeatedly" (see re-) + putare "to judge, suppose, believe, suspect" (originally "to clean, trim, prune," from PIE root *pau- (2) "to cut, strike, stamp"). Related: Rated; rating.

Old French reter also was borrowed into Middle English as retten "to blame" (c. 1300); also "to attribute, impute" (late 14c.), "to consider, think about" (late 14c.).

"estimate the worth or value of, reckon by comparative estimation," mid-15c., raten, from rate (n.). Intransitive sense of "have a certain value, rank, or standing" is from 1809; specifically as "have high value" by 1928. Related: Rated; rating.

Frame Rate - Frame rate, most commonly expressed in frames per second or FPS, is typically the frequency (rate) at which consecutive images (frames) are captured or displayed. This definition applies to film and video cameras, computer animation, and motion capture systems. In these contexts, frame rate may be used interchangeably with frame frequency and refresh rate, which are expressed in hertz. The temporal sensitivity and resolution of human vision varies depending on the type and characteristics of visual stimulus, and it differs between individuals. **The human visual system can process 10 to 12 images per second and perceive them individually, while higher rates are perceived as motion**. Modulated light (such as a computer display) is perceived as stable by the majority of participants in studies when the rate is higher than 50 Hz. This perception of modulated light as steady is known as the flicker fusion threshold. However, when the modulated light is non-uniform and contains an image, the flicker fusion threshold can be much higher, in the hundreds of hertz. With regard to image recognition, people have been found to recognize a specific image in an unbroken series of different images, each of which lasts as little as 13 milliseconds. Persistence of vision sometimes accounts for very short single-millisecond visual stimulus having a perceived duration of between

100 ms and 400 ms. Multiple stimuli that are very short are sometimes perceived as a single stimulus, such as a 10 ms green flash of light immediately followed by a 10 ms red flash of light perceived as a single yellow flash of light.

If it is not moving then it is ____. If a thing is not moving, and we make it move, what is that called? If I have something that is not moving and I make it start moving what is that called?

Does movement exist outside our mind? If movement light and sound only exist in or mind, what are the rules for inquiring into nature and for knowing all that exists, every mystery, every secret? Math?

- THE RHIND MATHEMATICAL PAPYRUS

God in the Bible is the number 26. Twenty six is the only number that exist between a square and a cube. 5 squared and 3 cubed, in that "space" or that space itself, is God. The space between a square and a cube, is the bridge, doorway, portal or vortex from 2 dimensions into 3 dimensions. We still good? Bwahahahahaha... 26 is double 13. If you think I am going to far, you have a long way to go. I promise you will grow more brain cells doing this type of abstract study, neurogenesis than study Diddy case or Kendrick beef! These are parables for you.

Should I double it (13) or halve it (26)? Egyptian

math, is a system of doubling and halving, this is how they understood the universe. Would you like another set to contemplate?

Set - Middle English setten, from Old English settan (transitive) "cause to sit; make or cause to rest as on a seat; cause to be put, placed, or seated;" also "put in a definite place," also "arrange, fix adjust; fix or appoint (a time) for some affair or transaction," and "cause (thoughts, affections) to dwell on."

This is from Proto-Germanic *(bi)satejanan "to cause to sit, set" (source also of Old Norse setja, Swedish sätta, Old Saxon settian, Old Frisian setta, Dutch zetten, German setzen, Gothic satjan), causative form of PIE *sod-, a variant of the root *sed- (1) "to sit." Also see set (n.2). It has been confused with sit (v.) at least since early 14c.

The intransitive sense of "be seated" is from c. 1200; that of "sink down, descend, decline toward and pass below the horizon" (of the sun, moon, or stars) is by mid-13c., perhaps from similar use of the cognates in Scandinavian languages; figurative use of this is from c. 1600.

Many uses are highly idiomatic, the verb, like put, its nearest equivalent, and do, make, get, etc., having become of almost universal application, and taking its distinctive color from the context. [Century Dictionary]

The sense of "make or cause to do, act, or be; start, bring (something) to a certain state" (on fire, in order, etc.) and that of "mount a gemstone" are attested by mid-13c. That of "determine upon, resolve" is from c. 1300; hence be set against "resisting" (mid-14c.).

The sense of "make a table ready for a meal" is from late 14c. (originally "set a board on trestles to serve as a dining table"); that of "regulate or adjust by a standard" (of a clock, etc.) also is from late 14c.

In printing, "to place (types) in the proper order for reading; put into type," 1520s. From c. 1500 as "put words to music." From 1570s as "put (a broken or dislocated bone) in position." In cookery, plastering, etc., "become firm or solid in consistency" by 1736.

To set (one's) heart on (something) is from c. 1300 as "love, be devoted to;" c. 1400 as "have a desire for." To set (one's) mind is from mid-15c.; transitive set (one's mind) to "determine to accomplish" is from late 15c. To set (something) on "incite to attack" (c. 1300) originally was in reference to hounds and game. To set an example is mid-14c. (set (v.) in the sense of "present" is from late Old English). The notion of "fix the value of" is behind old phrases such as set at naught "regard as nothing."

To set out is from c. 1300 as "display (for sale);" to set up shop "commence doing business" is from

c. 1400.

late Old English, sett, "appointed or prescribed beforehand;" hence "fixed, immovable, definite;" c.1300, of a task, etc., "imposed, prescribed;" past participle of setten "to set" (see set (v.)). By early 14c. as "ready." By 14c. with adverbs, "having a (specified) position, disposition, etc.;" by late 14c. as "placed, positioned;" to be set "be ready"

By 1510s as "formal, regular, in due form, deliberate;" 1530s as "placed in a setting, mounted." By c. 1600, of phrases, expression, etc., "composed, not spontaneous" (hence set speech, one planned carefully beforehand). By 1810 of the teeth, "clenched." The meaning "ready, prepared" is recorded from 1844.

By 1844 in reference to athletes poised to start a race, etc., or their muscles, "have or assume a rigid attitude or state." The exact phrase Get set! in the procedure of sprinting (after on your marks) is attested by 1890. A set piece, in theater, is "piece of free-standing scenery only moderately high, representing a single feature (such as a tree) and permitting more distant pieces to be seen over it" (by 1859); also, in the arts, "a painted or sculptured group" (1846).

"collection of matching things," mid-15c., sette, sete, earlier "religious sect" (late 14c.), in part from Middle English set, past participle of setten (see set (v.)) and in part from

Old French sette, sete "sequence," a variant of secte "religious community," from Medieval Latin secta "retinue," from Latin secta "a following" (see sect).

Skeat first proposed that set (n.), in the sense of "a number of things or persons belonging together" ultimately was a corruption of the source of sect, influenced by set (v.) in subsequent developments as if meaning "a number set together." Thus this noun set was in Middle English earliest in the sense of "religious sect," which also likely developed some modern meanings, such as "group of people" (mid-15c.), especially "persons customarily or officially associated" (1680s); "group of persons with shared status, habits, or affinities" (1777).

The meaning "a number of things having a resemblance or natural affinity; complete collection of pieces to be used together" is by 1560s. Hence, "collection of volumes by one author" (1590s), "complete apparatus for some purpose" (1891, of telephones, radio, etc.).

Meaning "group of pieces musicians perform at a club during 45 minutes" (more or less) is from c. 1925, though it is found in a similar sense from 1580s. Set-piece is from 1846 as "grouping of people in a work of visual art;" from 1932 in reference to literary works.

The word sett is a variant, preserved in old law and "now prevalent in many technical senses" [OED].

Egyptian god, from Greek Seth, from Egyptian Setesh.

"act of setting; state or condition of being set" (originally of the sun or another heavenly body), mid-14c., from set (v.) or its identical past participle. Old English had set "seat," in plural "camp; stable," but OED finds it "doubtful whether this survived beyond OE." Compare set (n.1).

Disparate senses collect under this word because of the many meanings given the verb. The sense of "manner or position in which something is set" is by 1530s, hence "general movement, direction, drift, tendency, inclination" (of mind, character, policy, etc.), by 1560s.

The meaning "permanent change of shape caused by pressure; a bend, warp, kink" is by 1812; that of "action of hardening," by 1837. Hence "action or result of fixing the hair when damp so that it holds the desired style" (1933).

"Something that has been set" (1510s), hence the use in tennis, "set of six games which counts as a unit" (1570s) and set-point "state of the game at which one side or player needs only one point to win the set" (by 1928).

The theatrical meaning "scenery for an individual scene in a play, etc.," is by 1859, from the past-participle adjective. It later was extended in movie and television production to the place or area where filming takes place.

Set (n.1) and set (n.2) are not always distinguished in dictionaries; OED has them as two entries, Century Dictionary as one. The difference of opinion seems to be whether the set meaning "group, grouping" (here (n.2)) is a borrowing of the unrelated French word that sounds like the native English one, or a borrowing of the sense only, which was absorbed into the English word.

Set Theory - Set theory is the branch of mathematical logic that studies sets, which can be informally described as collections of objects (number essences). Although objects of any kind can be collected into a set, set theory — as a branch of mathematics — is mostly concerned with those that are relevant to mathematics as a whole.

What are the paradoxes of set theory. As with most mathematical paradoxes, they generally reveal surprising and counter-intuitive mathematical results, rather than actual logical contradictions within modern axiomatic set theory. What is Russell's Paradox moor importantly? Russell's paradox shows that every set theory that contains an unrestricted comprehension principle leads to contradictions. According to the unrestricted comprehension principle, for any sufficiently well-defined property, there is the set of all and only the objects that have that property. Let R be the set of all sets that are not members of themselves. (This set is sometimes called "the Russell set".) If R is not a member of itself, then its definition entails that it is a member of itself; yet, if it is a member of itself, then it is not a member of itself, since it is the set of all sets that are not members of themselves. Don't think these "atheist' folks that are giving you, your scientific rules are studying anything else besides the Bible, Egypt & Sumer... Bwahahahahaha

This leads me to the number 137 or 1/137, numbers especially irregular numbers which reflect life (pun intended light = life, reflected get it), are abstract sets, codes within codes. 137 is the 33rd prime number.

- 1/137 was once thought to be the exact value of the fine-structure constant. The fine-structure constant, a dimensionless physical constant, is approximately 1/137, and the astronomer Arthur Eddington conjectured in 1929 that its reciprocal was in fact precisely the integer 137, which he claimed could be "obtained by pure deduction". This conjecture was not widely adopted, and by the 1940s, the experimental values for the constant were clearly inconsistent with the conjecture, being roughly 1/137.036. In 2021, researchers at the Kastler Brossel Laboratory in Paris reported the most precise measurement yet, determining the value to be 137.035999206 with an accuracy of 81 parts per trillion.

· Physicist Leon M. Lederman numbered his home near Fermilab 137 based on the significance of the number to those in his profession. Lederman expounded on the significance of the number in his 1993 book The God Particle: If the Universe Is the Answer, What Is the Question?, noting that not only was it the inverse of the fine-structure constant, but was also related to the probability that an electron will emit or absorb a photon—i.e., Feynman's conjecture. He added that it also "**contains the crux of electromagnetism (the electron), relativity (the velocity of light), and quantum theory (the Planck constant). It would be less unsettling if the relationship between all these important concepts turned out to be one or three or maybe a multiple of pi. But 137**?" The number 137, according to Lederman, "shows up naked all over the place", meaning that scientists on any planet in the universe using whatever units they have for charge or speed, and whatever their version of the Planck constant may be, will all come up with 137, because it is a pure number. Lederman recalled that Richard Feynman had even suggested that all physicists put a sign in their offices with the number 137 to remind them of just how much they do not know.

You ready for this? The Great Pyramid unites the numbers 26, 33 and 137. The Pyramid is 137 Meters High... 2.6 million Oblong Squares, built on the 33rd parallel! That's just a co-inky dink right? King Solomon's Temple brings Bible Study and Kemetic Study together! Solomon's Temple is a scale model of the Great Pyramid at 13.7 meters (18 to 21 inches per cubit, 30 cubits would be 540 to 630 inches, or 45 to 52.5 feet, or 13.7 to 16 meters, tall). Lets keep going...

137 is the most important age in the Bible. Ishmael, the son of Abraham and his wife Sarah's maidservant named Hagar, lived 137 years. Levi, the third son of the patriarch Jacob, had the same lifespan. Amram, the father of Moses and Aaron, additionally lived to the same age.

Genesis 25:17

And these are the years of the life of Ishmael, an hundred and thirty and seven years: and he gave up the ghost and died; and was gathered unto his people.

Exodus 6:16

And these are the names of the sons of Levi according to their generations; Gershon, and Kohath, and Merari: and the years of the life of Levi were an hundred thirty and seven years.

Exodus 6:20

And Amram took him Jochebed his father's sister to wife; and she bare him Aaron and Moses: and the years of the life of Amram were an hundred and thirty and seven years.

Remember we detailed the Jesus is a Naggar thing ie... Mark 6:3 etc...

Serpents are Naggars, serpents are the ancient symbols for being wise (Chakam).

Matthew 10:16

Behold, I send you forth as sheep in the midst of wolves: be ye therefore wise as serpents, and harmless as doves.

Broooooo Jesus is telling the 12 disciples to be wise as serpents, clearly he isn't suggesting they go behave as devils, so knock that devil talk off! Here for context...

Matthew 10

1 And when he had called unto him his twelve disciples, he gave them power against unclean spirits, to cast them out, and to heal all manner of sickness and all manner of disease.

2 Now the names of the twelve apostles are these; The first, Simon, who is called Peter, and Andrew his brother; James the son of Zebedee, and John his

brother;

3 Philip, and Bartholomew; Thomas, and Matthew the publican; James the son of Alphaeus, and Lebbaeus, whose surname was Thaddaeus;

4 Simon the Canaanite, and Judas Iscariot, who also betrayed him.

5 These twelve Jesus sent forth, and commanded them, saying, Go not into the way of the Gentiles, and into any city of the Samaritans enter ye not:

6 But go rather to the lost sheep of the house of Israel.

7 And as ye go, preach, saying, The kingdom of heaven is at hand.

8 Heal the sick, cleanse the lepers, raise the dead, cast out devils: freely ye have received, freely give.

9 Provide neither gold, nor silver, nor brass in your purses,

10 Nor scrip for your journey, neither two coats, neither shoes, nor yet staves: for the workman is worthy of his meat.

11 And into whatsoever city or town ye shall enter, enquire who in it is worthy; and there abide till ye go thence.

12 And when ye come into an house, salute it.

13 And if the house be worthy, let your peace come upon it: but if it be not worthy, let your peace return to you.

14 And whosoever shall not receive you, nor hear your words, when ye depart out of that house or city, shake off the dust of your feet.

15 Verily I say unto you, It shall be more tolerable for the land of Sodom and Gomorrha in the day of judgment, than for that city.

16 Behold, I send you forth as sheep in the midst of wolves: **be ye therefore wise as serpents**, and harmless as doves.

The point to our current convo is.... Drum roll please.

The Hebrew word chakam (Strong's Concordance #H2450) is recorded 137 times in 133 Hebrew Old Testament verses. It is written the most in Proverbs (46 times) followed by Ecclesiastes (21) and Jeremiah (11).

Proverbs 1:5
A wise man will hear, and will increase learning; and a man of understanding shall attain unto wise counsels:
Context?
Proverbs 1
1 The proverbs of Solomon the son of David, king of Israel;

2 To know wisdom and instruction; to perceive the words of understanding;

3 To receive the instruction of wisdom, justice, and judgment, and equity;

4 To give subtilty to the simple, to the young man knowledge and discretion.

5 A wise man will hear, and will increase learning; and a man of understanding shall attain unto wise counsels:

6 To understand a proverb, and the interpretation; the words of the wise, and their dark sayings.

7 The fear of the Lord is the beginning of knowledge: but fools despise wisdom and instruction.

Psalms 137

1 By the rivers of Babylon, there we sat down, yea, we wept, when we remembered Zion.

2 We hanged our harps upon the willows in the midst thereof.

3 For there they that carried us away captive required of us a song; and they that wasted us required of us mirth, saying, Sing us one of the songs of Zion.

4 How shall we sing the Lord's song in a strange land?

5 If I forget thee, O Jerusalem, let my right hand forget her cunning.

6 If I do not remember thee, let my tongue cleave to the roof of my mouth; if I prefer not Jerusalem above my chief joy.

7 Remember, O Lord, the children of Edom in the

day of Jerusalem; who said, Rase it, rase it, even to the foundation thereof.

8 O daughter of Babylon, who art to be destroyed; happy shall he be, that rewardeth thee as thou hast served us.

9 Happy shall he be, that taketh and dasheth thy little ones against the stones.

If we conservatively assume the cubit used by Noah to build the Ark was 18 inches (.45 meters), we end up with a ship that was 450 feet (137 meters) long, 75 feet (22.8 meters) wide, and 45 feet (13.7 meters) high.

Ready? Light and Sound only exist in your mind, Light and Sound can only even exist in your mind based on pigment content. The Flesh!!! Pigment literally changed the earth in a place where life could exist via oxygen. Chlorophyll the literally has 137 Atoms. To calculate the total number of atoms, we add up the number of each type of atom: 55 carbon atoms + 72 hydrogen atoms + 1 magnesium atom + 4 nitrogen atoms + 5 oxygen atoms = 137 atoms Therefore, there are 137 atoms in a molecule of chlorophyll A.

What we doing? I'm just saying, this is how I be standing on business when it come to the Bible! The Bible has infinite levels of 'Light'!
OK seriously I gotta wind this down…

The observer effect, do you have even more clarity

on how the observer effect works. Light only exists in your mind, beauty is in the eye of the beholder.

YOU HAVE TO HAVE ALL 3 OF THESE!!! PRAYER BOOKS

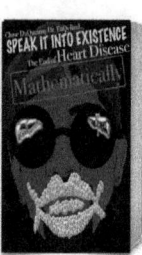  

THE RELATIONSHIP BETWEEN NUMEROSITY PERCEPTION AND MATHEMATICS ABILITY IN ADULTS: THE MODERATING ROLE OF DOTS

NUMBER

Ji Sun 1,2, Pei Sun 2,

Abstract

Background

It has been proposed that numerosity perception is the cognitive underpinning of mathematics ability. However, the existence of the association between numerosity perception and mathematics ability is still under debate, especially in adults. The present study examined the relationship between numerosity perception and mathematics ability and the moderating role of dots number (i.e., the numerosity of items in dot set) in adults.

Methods

Sixty-four adult participants from Anshun University completed behavioral measures that tested numerosity perception of small numbers and large numbers, mathematics ability, inhibition ability, visual-spatial memory, and set-switching ability.

Results

We found that numerosity perception of small numbers correlated significantly with mathematics ability after controlling the influence of inhibition ability, visual-spatial memory, and set-switching ability, but numerosity perception of large numbers was not related to mathematics ability in adults.

Conclusions

These findings suggest that the dots number moderates the relationship between numerosity perception and mathematics ability in adults and may contribute to explaining the contradictory findings in the previous literature about the link between numerosity perception and mathematics ability.

Keywords: Numerosity perception, Mathematics ability, Dots number, Small number, Large number, Numerosity discrimination

Introduction

Mathematics ability is a unique cognitive ability for humans, which is essential to our everyday life. Yet, there is a subgroup of people who have trouble acquiring mathematics skills. What factors influence mathematics ability? Previous studies have found that many domain-general factors could predict individual differences in mathematics ability, including memory (Raghubar, Barnes & Hecht, 2010), executive function (Kroesbergen et al., 2009) and early language ability (Hecht et al., 2001). In addition, symbolic number system (SNS) ability was an essential domain-specific predictor of mathematical ability (Lau et al., 2021; Lyons et al., 2018). For example, Geary & Vanmarle (2016) found that young children's core symbolic abilities significantly correlated with later mathematics achievement. Recently, several studies have found a potential role of numerosity perception in facilitating individuals' formal mathematical skills (e.g., Elliott et al., 2019; Halberda, Mazzocco & Feigenson, 2008; Mazzocco, Feigenson & Halberda, 2011; Park & Brannon, 2013). Furthermore, numerosity perception has been found to be a more important predictor of mathematics ability than other general cognitive functions (Chen & Li, 2014; Halberda, Mazzocco & Feigenson, 2008).

Numerosity perception is the elementary numerical ability in both humans and animals, representing nonverbal information of quantity without counting (Dehaene, 1997; Dehaene & Changeux, 1993; Feigenson, Dehaene & Spelke, 2004). A numerosity discrimination task is commonly applied to measure the precision of numerosity perception in which two dot sets are displayed briefly, and the participants are asked to judge which set is more numerous (Halberda & Feigenson, 2008; Inglis et al., 2011). The numerosity discrimination performance, such as weber fraction (w), was used as an index of numerosity perception. The w is an abstract measure of discrimination threshold that specifies the precision of the numerosity representation and represents the width of numerosity tuning curve (Kersey & Cantlon, 2017; Piazza et al., 2004), where higher w reflects less accurate numerosity representation.

The supporting evidence for the relationship between numerosity perception and mathematics ability came from the fact that numerosity discrimination performance (which reflects numerosity perception) correlated with mathematics ability (Halberda, Mazzocco & Feigenson, 2008; Libertus, Feigenson & Halberda, 2011; Mazzocco, Feigenson & Halberda, 2011), which was measured by different tasks (Holloway & Ansari, 2009; Fagerlin et al., 2007). Furthermore, it has been shown that training on numerosity perception tasks improved formal mathematics abilities in both children and adults (Hyde, Khanum & Spelke, 2014; Park & Brannon, 2013). However, it should be noted that many studies did not find such a relationship (Gilmore et al., 2013; Lyons & Beilock, 2011; Lyons et al., 2014). The inconsistent results might ascribe to two reasons. First, the inconsistency might be caused by age differences of participants recruited (Gilmore et al., 2013; Inglis et al., 2011; Schneider et al., 2017). For instance, Inglis et al. (2011) found that the association between numerosity perception and mathematics skills only existed in children but not in adults. Second, a methodological difference which refers to the different numerosity of items in the dot set (i.e., dots number) used in previous studies, might also contribute to the inconsistency. There is evidence that the number of stimuli that has to be estimated (i.e., dots number) could influence the perception mechanism (Anobile et al., 2016; Anobile, Cicchini & Burr, 2014; Anobile,

Cicchini & Burr, 2016; Pome et al., 2019; Fornaciai & Park, 2017; Zimmermann, 2018). It seems that small numbers of stimuli (relatively sparse dot-pattern) were sensed by the numerosity mechanism, but large numbers (relatively dense dot-pattern) were sensed by the density-texture mechanism (Anobile et al., 2016; Anobile, Cicchini & Burr, 2014; Zimmermann, 2018). Actually, previous studies found that only the numerosity perception of small numbers correlated with mathematics ability in children (Anobile et al., 2016).

Given the above, this study aimed to investigate the relationship between numerosity perception and mathematics ability and the effect of dots number on this relationship in adults. In particular, we conducted three measurements to assess participants 'mathematics ability, including an arithmetical test, a subjective mathematics ability evaluation, and their mathematics course scores. In the meantime, we conducted the numerosity discrimination tasks of small and large number conditions separately, which allowed us to evaluate the numerosity perception of different dot numbers. In addition, to control the potential effects of other general cognitive abilities on the relationship between numerosity perception and mathematics ability (Fuhs & McNeil, 2013; Gilmore et al., 2013), all participants performed three measurements on their inhibition ability, visual-spatial memory, and set-switching ability. We predicted that the number of dots would moderate the relationship between numerosity perception and mathematics ability in adults, and only the numerosity perception of small numbers would be related to mathematics ability.

Materials & Methods

Participants

Sixty-four undergraduate students from Anshun University (28 males and 36 females, Mage = 20.27, SDage = 1.11) participated in this study and received financial compensation for their time. This study was approved by the ethics committee of the School of Education Science, Anshun University (ID number: ASU-JYXY-201903) and all participants provided written informed consent at the beginning of the experiment. Three participants were excluded from the analyses due to either misunderstanding experimental task demand or missing data caused by computer errors.

Apparatus and software

All measurements were completed in a dimly lit room. Computer-based experiments were completed using MATLAB (R2016b; MathWorks, Cambridge, MA, USA) and PsychToolbox (Brainard, 1997) on a 23-inch LED monitor (Dell: U2312 HM) with 1,600 × 1,080 resolution at 60 Hz and a standard desk (width: 70 cm; length: 150 cm), viewed binocularly from a distance of 60 cm.

Measures

Numerosity perception, mathematics ability, inhibition ability, visual-spatial memory, and set-switching ability were measured in the following sequential order. The whole experimental procedure took approximately 85 min.

Numerosity perception

A modified version of a numerosity discrimination task adapted from Anobile, Cicchini & Burr (2014) was used to measure numerosity perception. Two sets of dots were of equal size and presented simultaneously for 250 ms on either left or right side of a central fixation point. Each set was constrained to a 14°-diameter virtual circle and comprised a number of 0.3° (in visual angle) diameter dots, half white and half black. For each trial, the set on the right side of the fixation point was the reference, and the left set was the probe. In separate blocks, the reference always contained 11 dots (small number condition) or 152 dots (large number condition), while the numerosity of the probe was changed from trial to trial, following the Method of Constant stimuli that varied the number of dots from 50% to 200% of the number of the reference that was split into seven equal log unit steps. For small number condition, the probe patches contained 6, 7, 9, 11, 14, 17, or 22 dots. For large number condition, the probe patches contained 76, 96, 121, 152, 192, 241, or 304 dots. Each of the seven probe stimulus levels was tested for 16 trials, and each block consisted of 112 trials. For each trial, participants were asked to judge as accurately and quickly as possible which set had more dots. Two blocks were presented in counterbalanced order. Participants gave their responses using a keyboard. Based on the pilot study, 14 practice trials were given to participants with feedback before the formal data collection.

"The proportion of probe greater trials" was plotted against the reference number and fitted with the Methods of Probits (Sun et al., 2013). The 50% point estimated the point of subjective equality (PSE), and the difference in numerosity between the 25% and 75% points provided the just notable difference (JND), which was then divided by PSE to estimate the Weber fraction (w).

Mathematics ability

Three measures were used to assess participants' mathematics ability, including a modified-version arithmetical test (AT), a subjective mathematics ability evaluation, and the mathematics course score. The arithmetical test, which was designed based on Shalev et al. (2001), assessed basic arithmetic skills. The types of items that were tested included number facts, complex arithmetic, decimals calculation, and fraction calculation. In the self-developed measure of subjective mathematics ability evaluation, participants were instructed to evaluate their mathematics ability compared with other people using a 10-point scale (1 = the worst level, 10 = the best level). Furthermore, participants 'college entrance examination score was used as the mathematics course score. Following the method of convention of Holloway & Ansari (2009), the scores for the three tasks were converted into z-scores and averaged to create an overall mathematics ability score for each participant, which was then used in the following analyses as an index of participants' mathematics ability.

Inhibition ability

Inhibition ability is the ability to suppress dominant, automatic, or prepotent response for irrelevant or no-longer-relevant information (Toll et al., 2011). The classical Go/No-go task was used to assess inhibition ability (Falkenstein, Hoormann & Hohnsbein, 1999). In this task, participants were told to respond to a yellow square (Go stimuli) and stop responding to a yellow circle (No-go stimuli). Participants were given five practice trials with feedback. The official task had 200 test trials without feedback, including 150 Go trials and 50 No-go trials. For each participant, the accuracy for No-go trials was recorded as the inhibition score.

Visual-spatial memory

Visual-spatial memory is a system for retaining location and object information (Wood, 2011). To assess participants 'visual-spatial memory ability, a widely used memory span task for visual patterns was administered (Mejias, Gregoire & Noel, 2012). The task was developed by Wilson, JH & Power (1987) and consisted of the presentation of boxes with some boxes being filled in. The participants were instructed to remember the pattern of the boxes, which was displayed for 2 s. After a blank interval of 2 s, the boxes would be presented again with one box missing. The participants had to recall which box was missing. The initial pattern involved filling in two boxes, and upon receiving two correct responses out of three attempts, the complexity of the pattern would increase. The final number of successfully recalled boxes was recorded as a memory span score.

Set-switching ability

Set-switching ability is the ability to direct actions and thoughts to selected goals; it has been thought of as a kind of executive control. In this study, a modified Trail Making Test was presented in the paper-and-pencil version to measure set-switching ability (Arbuthnott & Frank, 2000). The test consisted of three trials: Part A-1, Part A-2, and Part B. Part A-1 included drawing a line connecting consecutive numbers from 1 to 25. Part A-2 included drawing a line connecting consecutive letters from A to Y. Part B included drawing a line connecting alternating numbers and letters in sequence (i.e., 1-A-2-B and so on). Participants were shown three trials in order. In each trial, a timer was set when the participants began drawing a line on paper and stopped when they put the pencil on the desk. The time to complete each trial was recorded. The ratio of completion time between Part B and Part A (average completion time of Part A-1 and Part A-2) was calculated as a set-switching score. A higher set-switching score indicated a poorer set-switching ability.

Data analysis

The statistical analyses of data were performed using SPSS software (Version 23.0). The correlations between all variables were measured using Zero-order correlation analyses. Then, we used hierarchical regression analysis to explore the predictors of mathematics ability.

Results

Mathematics ability score

We looked for the correlations of the participants' performance on the three mathematics ability tests. Subjective mathematics ability evaluation was significantly correlated with AT accuracy ($r = 0.32$, $p = 0.013$) and the mathematics course score ($r = 0.32$, $p = 0.012$). However, AT accuracy was not significantly correlated with mathematics course scores, $r = 0.15$, $p = 0.257$. Given that mathematics ability is a complex construct and includes different skills (Mannamaa et al., 2012), these results showed that performances in three tests measured not only common aspects but also different aspects of mathematics ability. This might, in part, explain the previous contradictory relationships between numerosity perception and mathematics ability. To provide an evaluation of overall mathematics ability, we followed Holloway

& Ansari (2009) and used the mathematics ability score for each participant in further analyses.

The correlations between numerosity perception and mathematics ability

A zero-order correlation matrix was calculated to investigate the relationship between all measurements (See Table 1). Results showed that mathematics ability significantly correlated with wsmall ($r = -0.42$, $p = 0.001$; the alternative Pearson $r = -0.36$, $p = 0.004$, see Fig. 1), but not with wlarge ($r = 0.10$, $p = 0.444$; the alternative Pearson $r = -0.11$, $p = 0.414$), indicating that lower mathematical ability was associated with higher w (reflecting poorer numerosity discrimination precision) in the small number condition. Further correlation analysis showed that wsmall was significantly correlated with AT accuracy and the mathematics course score ($r = -0.32$, $p = 0.017$; $r = -0.37$, $p = 0.005$), but not with subjective mathematics ability evaluation ($r = -0.21$, $p = 0.125$). Conversely, we found no significant correlations between wlarge and three mathematics ability tests, $rs < 0.24$, $ps > 0.075$. It should be noted that there was also a significant correlation between wsmall and wlarge ($r = 0.47$, $p < 0.001$; the alternative Pearson $r = 0.53$, $p < 0.001$). Here, we further examined the unique effect of dots number (i.e., small number and large number conditions) on the relationship between numerosity perception and mathematics ability. We used a specific test of the difference between two dependent correlations with one variable in common developed by Lee & Preacher (2013) to compare the correlations between wsmall, wlarge and mathematics ability. Results showed that there was a significant difference between the correlation of wsmall and mathematics ability and the correlation of wlarge and mathematics ability, $p < 0.001$. These results indicated the dots number moderated the relationship between numerosity perception and mathematics ability.

THE NEURODEVELOP MENTAL BASIS OF MATH ANXIETY

Christina B Young 1, Sarah S Wu 1, Vinod Menon

Abstract

Math anxiety is a negative emotional reaction to situations involving mathematical problem solving. Math anxiety has a detrimental impact on an individual's long-term professional success, but its neurodevelopmental origins are unknown. In a functional MRI study on 7- to 9-year-old children, we showed that math anxiety was associated with hyperactivity in right **amygdala** regions that are important for processing negative emotions. In addition, we found that math anxiety was associated with reduced activity in posterior parietal and dorsolateral prefrontal cortex regions involved in mathematical reasoning. Multivariate classification analysis revealed distinct multivoxel activity patterns, which were independent of overall activation levels in the right **amygdala**. Furthermore, effective connectivity between the **amygdala** and ventromedial prefrontal cortex regions that regulate negative emotions was elevated in children with math anxiety. These effects were specific to math anxiety and unrelated to general anxiety, intelligence, working memory, or reading ability. Our study identified the neural correlates of math anxiety for the first time, and our findings have significant implications for its early identification and treatment.

Keywords: cognitive neuroscience, cognitive development, educational psychology, neuroimaging, mathematical ability

Strong mathematical skills are increasingly essential for academic and professional success in today's high-technology world (National Mathematics Advisory Panel, 2008). Research has shown that math anxiety has a negative impact on mathematical skills, which leads to adverse effects on career choice, employment, and professional success (Ma, 1999). Math anxiety is thought to influence learning and mastery of mathematics from an early age, but its precise developmental origins are not known (Rubinsten & Tannock, 2010). Although the first years of elementary schooling are an important period for acquiring basic mathematical skills, previous behavioral studies of math anxiety have mainly focused on adolescents and adults. However, across all age groups, but most notably in children, nothing is currently known about the neurobiological mechanisms underlying math anxiety. The study reported here is the first to identify the neural basis of math anxiety in young children and demonstrate its impact on brain functioning and connectivity at one of the earliest stages of formal acquisition of math skills.

Math anxiety is a negative emotional response that is characterized by avoidance as well as feelings of stress and anxiety in situations involving mathematical reasoning (Ashcraft & Ridley, 2005). It can often hinder the successful completion of tasks involving manipulation of numerical information and is a prominent cause of problem-solving difficulties across all age ranges (Ashcraft & Krause, 2007; Suinn, Taylor, & Edwards, 1988; Wigfield & Meece, 1988). Behavioral studies, mainly in adults, have shown that math anxiety has a negative effect on performance of basic numerical operations, including counting, addition, and subtraction (Ashcraft & Ridley, 2005; Maloney, Risko, Ansari, & Fugelsang, 2010). Because the detrimental impact of math anxiety on mathematical development is lifelong (Bynner & Parsons, 1997; Rubinsten & Tannock, 2010), it is important to understand its neurobiological origins, especially during the earliest stages of formal math learning in elementary school children.

Although the behavioral literature on math anxiety in adults and adolescents is extensive, there is a relative dearth of studies investigating math anxiety in young children. This is in part due to the lack of a developmentally appropriate measure of math anxiety. To address this issue, we recently extended the Mathematics Anxiety Rating Scale (MARS), a standardized method for assessing math anxiety in older children and adults (Richardson & Suinn, 1972; Suinn & Edwards, 1982), to create the Scale for Early Mathematics Anxiety (SEMA; Table S1 in the Supplemental Material available online provides details of this scale). The SEMA has been shown to be a reliable and construct-valid (Cronbach's α = .870) measure of math anxiety in 7- to 9-year-old second and third graders (Wu, Amin, Barth, Melcarne, & Menon, 2012).

To examine the neurodevelopmental basis of math anxiety, we analyzed functional brain-imaging data from forty-six 7- to 9-year-old children, which we obtained while the children determined whether addition and subtraction problems were correct (e.g., "2 + 5 = 7") or incorrect (e.g., "2 + 4 = 7"). In a separate session, we used the SEMA to assess math anxiety in each child. Previous studies in both children and adults have implicated multiple limbic, paralimbic and prefrontal cortex regions, including the **amygdala** and medial prefrontal cortex in social and generalized anxiety disorders (Etkin & Wager, 2007; Guyer et al., 2008). Normal healthy adults also show activation of these same regions while viewing negative stimuli, such as fearful faces (Sabatinelli et al., 2011). However, it is not known whether these same areas are also engaged by "neutral" numerical symbols that are perceived negatively. Specifically, it is unknown whether these limbic-prefrontal cortex circuits are differentially engaged during math problem solving in individuals with high math anxiety. We hypothesized that if children with high math anxiety view such stimuli negatively, they would show hyperactive **amygdala** response during math problem solving. Furthermore, **amygdala** connectivity with medial prefrontal cortex regions involved in emotion regulation would also be elevated when compared with such connectivity in children with low math anxiety.

We used multivariate pattern analysis (MPA; Kriegeskorte, Goebel, & Bandettini, 2006) to further investigate aberrant patterns of amygdala response in children with high math anxiety. We predicted that such children would show distinct spatial patterns of task-related activity in the **amygdala** when compared with children with low math anxiety and that the activation patterns of each group would be independent of overall differences in signal amplitude. Finally, we hypothesized that children with high math anxiety would show decreased engagement of the intraparietal sulcus and dorsolateral prefrontal cortex regions typically associated with mathematical cognition in children (Ansari, 2008; Rivera, Reiss, Eckert, & Menon, 2005).

Method
Participants
A total of 54 children from the San Francisco area were originally recruited for this study. Eight participants were excluded because of poor functional MRI (fMRI) data quality. The remaining sample (n = 46) consisted of 28 boys and 18 girls in the second and third grades (age range = 7.17–9.33 years, M = 8.42 years, SD = 0.61 years). All but 4 children were right-handed. None of the participants had a history of psychiatric illness, neurological disorders, or learning disabilities. Participants were recruited via flyers sent to elementary schools as well as advertisements posted in libraries, in magazines, on Web sites, and with learning-disability groups. All protocols were approved by the institutional review board at Stanford University School of Medicine, and participants were treated in accordance with the American Psychological Association Code of Conduct.

Neuropsychological assessments

Before the fMRI scan, we conducted a neuropsychological assessment on each child. This assessment consisted of the Wechsler Abbreviated Scale of Intelligence (Wechsler, 1999) to measure IQ, the Wechsler Individual Achievement Test (Wechsler, 2001) to assess performance, and the Working Memory Test Battery for Children (Pickering & Gathercole, 2001) to determine working memory capacity. Parents of study participants also completed the Child Behavior Checklist for Ages 6–18 (CBCL/6–18; Achenbach & Rescorla, 2001), and trait anxiety was determined using the DSM-Oriented Anxiety Problems subscale of the CBCL. The SEMA (Wu et al., 2012) was used to assess math anxiety. SEMA scores were used both as a continuous measure and to divide participants into high-math-anxiety (HMA) and low-math-anxiety (LMA) groups. All measures have been shown to have accurate validity and reliability and have been normed for use in children.

Functional brain imaging

Each child completed two runs in the fMRI scanner: the addition run and the subtraction run. Each run had four task conditions: (a) complex arithmetic problems, (b) simple arithmetic problems, (c) number identification, and (d) passive fixation. Only the two arithmetic-task conditions were examined in this study. In both arithmetic conditions, full equations were given, and the child indicated via a button box whether the answer shown was correct or incorrect (see fMRI Experimental Design in the Supplemental Material).

In the addition run, complex problems consisted of equations with one addend ranging from 2 to 9 and the other addend ranging from 2 to 5 (e.g., "5 + 2 = 7"). There were no problems in which both addends were the same (e.g., "5 + 5 = 10"). The format of simple addition problems was identical to that of complex problems, except that one of the addends was 1 (e.g., "5 + 1 = 7"). The design of the subtraction run was the same as its addition counterpart, such that complex subtraction problems were the inverse of complex addition problems (e.g., "7 – 5 = 2"), and simple subtraction problems contained a subtrahend of 1 (e.g., "7 – 1 = 5"). In each case, incorrect answers deviated by ±1 or ±2 from the correct answer. Critically, for both addition and subtraction, the complex and simple problems had equivalent numerical and symbolic formats as well as the same response-selection requirements.

fMRI data acquisition and analysis

Acquisition and preprocessing

Functional brain images were acquired on a 3-T GE Signa scanner (see fMRI Data Acquisition in the Supplemental Material). Data were analyzed using a general linear model implemented in the Statistical Parametric Mapping program (SPM8; Wellcome Trust Centre for Neuroimaging, London, United Kingdom). Images were realigned to correct for movement, denoised, spatially normalized to Montreal Neurological Institute (MNI) space, and smoothed with an effective Gaussian kernel of 6 mm (see fMRI Preprocessing in the Supplement Material). A repeated measures analysis of variance of movement parameters using group (HMA, LMA) as a between-subjects factor and operation (addition, subtraction) and direction (x, y, z) as within-subjects factors revealed that there were no group differences in movement, $F(1, 44) = 0.661$, $p = .420$. A linear regression analysis confirmed that SEMA scores were not correlated with movement ($p > .380$).

Statistical analysis

Data from the two runs were combined in a single analysis. Brain activity related to each task condition was modeled by convolving boxcar functions with a canonical hemodynamic response function and a temporal derivative to account for voxel-wise latency differences in hemodynamic response. Voxel-wise t statistics were generated for each participant by contrasting complex addition and subtraction problems with simple addition and subtraction problems. Brain responses in the HMA and LMA groups were then compared using a t test on contrast images from each participant. Significant clusters of activation were determined at a voxel-wise height threshold of $p < .01$, with family-wise error (FWE) correction for multiple spatial comparisons ($p < .01$, $k = 133$ voxels). The FWE correction was determined using Monte Carlo simulations (Ward, 2000).

In addition to conducting the dichotomous group analysis, we used SEMA scores as a continuous variable to identify brain regions that showed increases and decreases in brain activation related to math anxiety. Cytoarchitectonic maps (Eickhoff et al., 2005) were used to determine the percentage of fMRI activation clusters that fell within individual subdivisions of the **amygdala** and **anterior hippocampus** as well as to conduct (anatomically) unbiased MPA. MPA was used to examine whether multivoxel spatial activation patterns, above and beyond overall differences in signal level, were different between the two groups (see fMRI Multivariate Pattern Analysis in the Supplemental Material). Finally, psychophysiological interaction analysis was used to examine group differences in effective connectivity of the **amygdala** independent of overall task-related activation (see fMRI Effective Connectivity Analysis in the Supplemental Material).

Results
Math anxiety and behavior
SEMA scores were neither uniformly distributed, $\chi^2(4, N = 46) = 19.00$, $p = .001$, nor normally distributed (Shapiro-Wilk test $W = 0.91$, $p = .002$).

To increase sensitivity and robustness, and to facilitate interpretability and multivariate classification analysis of fMRI data, we used a median split on the total-anxiety score to divide children into two groups of equal size (HMA group score: M = 38.391, SD = 7.590; LMA group score: M = 25.348, SD = 2.673). The two groups differed on math anxiety but not on IQ, working memory, or overall reading and math proficiency (Table 1). Math-anxiety scores did not differ between boys and girls, $F(1, 44) = 1.751$, $p = .193$, or between second and third graders, $F(1, 44) = 0.018$, $p = .894$, so data from both genders and grades were pooled for subsequent analysis. Critically, although the two groups differed in math anxiety, $F(1, 44) = 60.425$, $p < .001$, they did not differ in trait anxiety, $F(1, 39) = 0.508$, $p = .480$. Additional analyses using math anxiety and trait anxiety as continuous variables showed that these two measures were not significantly correlated $(r = -.132, p > .400)$.

Full circle now get it? Circle, Pi, 22/7? Kemet and the Bible linked from the first lines of the Bible… 22/7 is exclusively the Kemetic approximation of Pi and after that it the Math of Creation the entire Bible is **based** on!

I love the Bible and all it's books and their Authors, even Paul!

Jesus the man is a rough stone, Christ is the

perfected stone. If you have to go to the Qumran Scrolls, the Bagvegita, the Jewish Sanhedrin, the Gnostic Pistis Sophia, the Koran, the Walls of Kemet, the Papyri, the Sumerian Tablets, the Torah to help you understand God, not the prophets, the messengers etc... but God, do that.

My Father was Catholic and they have 72 books, the Protestants use 66 books, respectfully the Bible so powerful that just a couple verses can change your life, inspire cities to be built or destroyed, build or destroy governments, civilizations etc... Don't sweat the small stuff, don't get tricked out of position either. There are many text, papyri, scriptures associated with the Bible, the more skeptical you are, the stronger your conviction will be, as your synderesis moves you once you finally understand the truth. There is only 1 Most High, he is in heaven, I hope you read this book and ascribe all it's flaws to me and all the Light to him!

GO TO
AMAZON.COM

YOU HAVE TO HAVE ALL 3 OF THESE!!!
PRAYER BOOKS

www.ingramcontent.com/pod-product-compliance
Lightning Source LLC
Chambersburg PA
CBHW071447220526
45472CB00003B/700